OUR AWESOME

Its Mysteries and Its Splendors

Prepared by the Special Publications Division
National Geographic Society, Washington, D.C.

EARTH

Our Awesome Earth:
Its Mysteries and Its Splendors

Contributing Authors: Chris Eckstrom Lee,
  Paul Martin, Jane R. McCauley, Thomas O'Neill,
  Cynthia Russ Ramsay

Published by The National Geographic Society
Gilbert M. Grosvenor, *President*
Melvin M. Payne, *Chairman of the Board*
Owen R. Anderson, *Executive Vice President*
Robert L. Breeden, *Senior Vice President,
  Publications and Educational Media*

Prepared by The Special Publications Division
Donald J. Crump, *Director*
Philip B. Silcott, *Associate Director*
Bonnie S. Lawrence, *Assistant Director*

Staff for this Book
Richard M. Crum, *Managing Editor*
John G. Agnone, *Illustrations Editor*
Jody Bolt, *Art Director*
Stephen J. Hubbard, *Project Coordinator
  and Senior Researcher*
Monique F. Einhorn, Tee Loftin, Lucinda Moore,
  Researchers
Paul Martin, H. Robert Morrison, Pamela Black
  Townsend, Suzanne Venino, *Picture Legend Writers*
Marianne R. Koszorus, *Assistant Art Director*
Pam Castaldi, *Assistant Designer*
Artemis S. Lampathakis, *Illustrations Assistant*
John D. Garst, Jr., *Director, Publications Art;*
  Virginia L. Baza, Isaac Ortiz, D. Mark Carlson,
  Daniel J. Ortiz, *Map Research and Production*

Engraving, Printing, and Product Manufacture
Robert W. Messer, *Manager*
David V. Showers, *Production Manager*
George J. Zeller, Jr., *Production Project Manager*
Gregory Storer, *Senior Assistant Production
  Manager*
Mark R. Dunlevy, *Assistant Production Manager*
Timothy H. Ewing, *Production Staff Assistant*
Mary F. Brennan, Vicki L. Broom, Carol Rocheleau
  Curtis, Lori E. Davie, Mary Elizabeth Davis, Ann Di
  Fiore, Rosamund Garner, Bernadette L. Grigonis,
  Virginia W. Hannasch, Nancy J. Harvey, Joan Hurst,
  Katherine R. Leitch, Ann E. Newman, Cleo E.
  Petroff, Stuart E. Pfitzinger, Virginia A. Williams,
  Staff Assistants
Dianne L. H. Hardy, *Indexer*

Library of Congress CIP Data: page 199.

*Attentive African lioness listens to a cub's meek snarl.
PRECEDING PAGES: Carrying supplies, Sherpas
trudge into the Himalayas, highest mountains on
earth. PAGE 1: Molten rock erupts from Hawaii's
Kilauea, one of the world's most active volcanoes.*

HARDCOVER: BUCK MOUNTAIN ETCHES THE SKY BENEATH A WYOMING MOON.

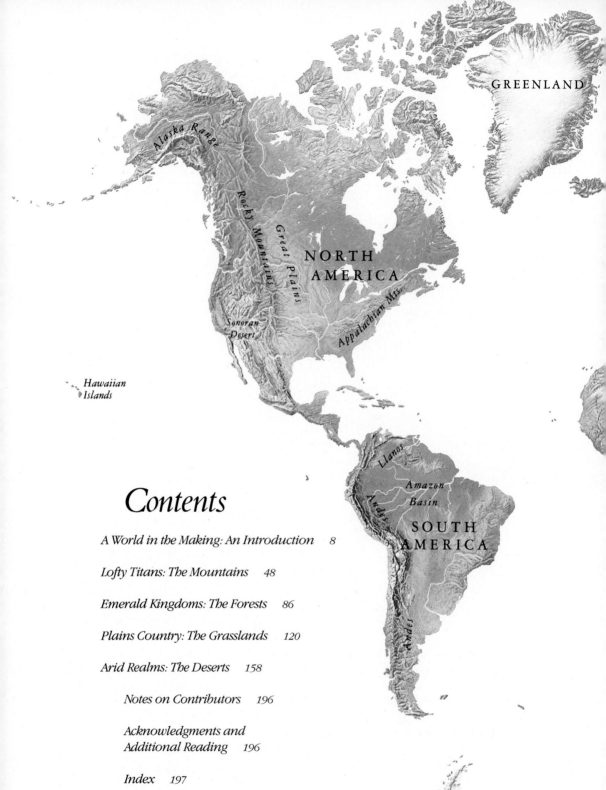

GREENLAND

Alaska Range

Rocky Mountains

Great Plains

NORTH AMERICA

Sonoran Desert

Appalachian Mts.

Hawaiian Islands

Llanos

Amazon Basin

Andes

SOUTH AMERICA

Andes

Contents

EUROPE

Alps

ASIA

Siberia

Ural Mts.

The Steppes

Gobi Desert

Sahara

Arabian Peninsula

Himalayas

Sudan

AFRICA

Serengeti Plain

Kalahari Desert

AUSTRALIA

Great Victoria Desert

Natural theaters of mystery and splendor: Deserts, forests, grasslands, and mountains dominate the continents. One-third of earth's landmass is desert. The largest cold desert, Antarctica locks up five million square miles of ice. The darker green shading shows the forests, ranging from equatorial belts to polar fringes. The lighter green hue traces the grasslands. Mostly under cultivation, they make up one-fourth of earth's land area. Thrust to grand heights by geologic bumping and wrenching, mountains crown each continent with rugged majesty.

MAP ILLUSTRATION BY
NATIONAL GEOGRAPHIC CARTOGRAPHIC DIVISION

ANTARCTICA

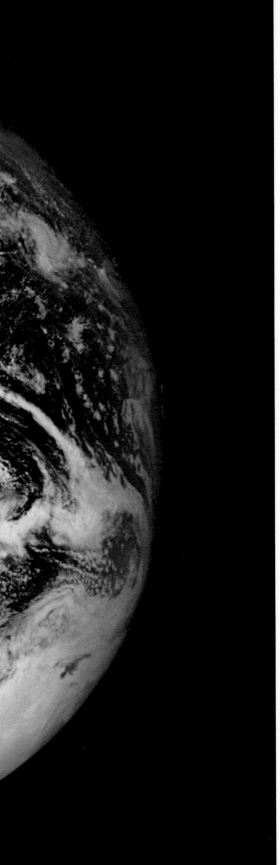

A World in the Making

An Introduction

By Paul Martin

Swirling clouds turn planet Earth into a shining agate. At upper left a desert belt extends from Africa across the Red Sea to the Arabian Peninsula. In the southernmost region the starkness of Antarctica blends with clouds that hook northward across the tip of Africa. Oceans cover most of the "blue marble" planet, a living sphere shared by 4.8 billion people and some 2 million known species of plants and animals.

FOLLOWING PAGES: Molten rock from Mauna Loa volcano in Hawaii affirms the power of our restless planet.

*T*enacious ama'uma'u fern (below) signals
the advance of an army of colonizing
plants on a lava flow in Hawaii Volcanoes
National Park. In dry areas, centuries-old
lava often remains devoid of vegetation,
but with sufficient rainfall, plants return to
eruption sites in a matter of months. In
time, moist areas may regain the lushness
of the mature Hawaiian rain forest at right.
Here two of the islands' most common
plants, luxuriant hapu'u ferns in the
foreground and gnarled 'ohi'a trees,
flourish in rich volcanic soil. In ideal
settings, 'ohi'a trees soar to the height of an
eight-story building, and hapu'u ferns lift
their delicate fronds 25 feet into the air.
Evolving in isolation—2,000 miles from
Asia and North America—Hawaii's native
flora once numbered about 1,300 species.
Since Capt. James Cook's discovery of the
islands in 1778, some 2,000 new plants
have been introduced, often to the
detriment of native species. Amid these isles
of haunting beauty—as elsewhere around
the globe—change comes as dramatically
as a volcanic eruption or as quietly as a
sprouting plant.

PAUL CHESLEY (ABOVE); DAVID MUENCH (RIGHT)

An Introduction

ore than a century ago, naturalist John Muir contemplated the earth as "one great dewdrop, striped and dotted with continents and islands, flying through space." Today, that poetic vision has been given substance. Photographs taken from space reflect a shimmering portrait of Muir's "great dewdrop," capturing the subtle colors and textures of the planet's dominant features—its seas, mountains, forests, grasslands, and deserts. The authors of this book set out to explore those varied domains. Among their destinations: the volcanoes of Indonesia, the cloud forests of Costa Rica, the Serengeti grasslands, and the vast Sahara.

The authors' travels revealed much about earth's major biomes, each with an array of plants and animals remarkably adapted to their environment. In her studies of mountains, Cynthia Ramsay learned of the incredible chamois, a European goatlike animal that nimbly bounds across 20-foot chasms. As Jane McCauley explored the world's forests, she discovered how trees thrive in the Amazon's infertile soil by voraciously extracting nutrients from decaying plants. Investigating earth's grasslands, Chris Lee encountered the world's largest herd of migrating animals—Africa's wildebeest. In his desert explorations, Tom O'Neill came upon a world of unexpected life that included the hardy Sonoran desert lily, which can survive years of drought—blossoming only when aroused by sufficient rains.

The authors also encountered a wealth of mysteries about our planet. How are the Sherpas of Nepal able to extract adequate amounts of oxygen from the air of the Himalayas, which for most people is too thin for comfortable breathing? What causes the tundra-like openings called heath balds in the forests of the Great Smokies? What secrets do elephants share in tones too low for human ears to pick up? What is the significance of Peru's Nazca Lines, huge figures and geometric designs scraped on the desert floor perhaps 2,000 years ago?

Even more basic questions arose. How were the world's mountains thrust up to such heights? What determines whether an area becomes forest or grassland or desert? What unimaginable forces move and sculpt the continents? I considered these and other questions as I explored the ocean realm and lands edged by the sea. There, I found that answers to some of earth's riddles are penned in a fiery scrawl.

I am walking through a tropical Eden on the edge of creation. About me grows a dense rain forest, dripping with moisture, hushed but for the soft trill of birds. Yet only steps away lies a landscape as barren as any on earth. I am hiking the Crater Rim Trail in Hawaii Volcanoes National Park, on the island of Hawaii. From overlooks along the trail I can gaze across the two-mile-wide summit crater, or caldera, of one of earth's most active volcanoes, Kilauea.

Sheer walls drop 400 feet to the caldera's rolling floor of hardened lava. Wisps of steam form airy flags that hint at the pent-up heat of magma—molten rock—seething more than a mile below the crust. I have come to the Big Island to see a part of the earth recently renewed by fire. Kilauea has been erupting regularly, 40 episodes in 36 months since the current series began in January 1983.

In the past decade alone, Hawaii's two active volcanoes, Kilauea and Mauna

Loa, have spewed up billions of cubic yards of island-building lava. Measured from seafloor to summit, Mauna Loa rises nearly 32,000 feet, the greatest mountain mass on earth. Hawaii Volcanoes National Park encompasses portions of both Mauna Loa and Kilauea, providing a refuge for many native plants and animals.

Crater Rim Trail leads me through a lushness typical of the Hawaiian chain. From this forest paradise, I follow the trail down into a pit of desolation, the crater of Kilauea Iki, or Little Kilauea. This mile-long crater lies just to the east of the summit. Alone, I cross a landscape conceived in a demented dream. The crater floor seems to writhe with contorted shapes: Frozen waves of lava pitch about me; giant heaps of slag tumble and slide; massive slabs tilt like reeling drunkards.

"This is a place that gives meaning to the word awesome," chief naturalist Tom White tells me when I return to park headquarters. "It's a place where the power of nature is always on display." Days later, Kilauea exhibits its power anew as a pulsing orange-hot fountain of lava bursts from Pu'u O'o, a cinder cone on the volcano's eastern flank. Late in the evening the fountain peaks at just over a thousand feet—a jet twice the height of the Washington Monument. Millions of cubic yards of lava pour down the mountain. Trees that are engulfed burst into flame like candles on some immense, orange-frosted cake.

Early the following morning the fountaining stops. Soon after the eruption subsides I fly over Pu'u O'o and stare down into an open artery leading from earth's pulsing heart. A searing pool of molten rock still simmers inside the cinder cone. Our tiny aircraft banks low over the cone at just above stall speed. Turbulence from the updraft shakes the plane like a child's kite. As I open the side window for a clearer view, the volcano's heat blasts my face, filling my nostrils with the stench of brimstone. We cruise over the glowing lava flow to gauge its extent. The leading edge runs some three miles from Pu'u O'o. The flow has crept dangerously close to the Royal Gardens Subdivision, where several homes have been destroyed by previous eruptions. Atop one house I see a grimly humorous sign: "Hot Lots For Sale—Real Cheap."

Assessing the potential destructiveness of these earth-shaping rivers of fire is a task of the Hawaiian Volcano Observatory, located in the park on the rim of Kilauea caldera. Established in 1912, the observatory is part of the U.S. Geological Survey's volcano investigation program. The scientist in charge is Thomas L. Wright. He tells me that the Big Island's volcanic history is but the current passage in an age-old geologic tale. The most startling part of this story is the likelihood that the entire Hawaiian archipelago, along with the 2,000-mile string of underwater mountains called the Emperor Seamounts, was forged by the very fires now fueling Mauna Loa and Kilauea.

Explains Dr. Wright, "Each of the islands in the Hawaiian chain formed above a single, fixed hot spot, an area where a jet or plume of magma from within the earth forces its way upward through the planet's crust. One at a time, the Hawaiian Islands rose from the ocean floor over millions of years."

Yet how can a stationary source of magma account for a chain of islands and seamounts extending across 4,000 miles of ocean? The apparent answer to this riddle lies in the theory of plate tectonics. According to this widely accepted concept, the earth's crust is made up of several large rigid plates, along with many smaller ones. Like colossal rafts, these plates slowly float about on a layer of partially molten material. Thomas Wright's predecessor at the observatory, Robert Decker, has studied and written about this revolutionary theory.

"In the early 1900s a German meteorologist named Alfred Wegener proposed the idea that North and South America had broken away from Europe and Africa, and had slowly drifted apart," explain Dr. Decker and his coauthor and wife, Barbara. "He championed this idea for 20 years with attractive arguments like the jigsaw-puzzle fit of the continents bordering the Atlantic Ocean. He was so zealous that many scientists considered him a crackpot. When geologists determined that the rocks of the ocean floor were too strong to allow the continents to move slowly through them, his ideas were discredited."

P rior to World War II, however, new oceanographic findings began to vindicate Wegener. By 1960, scientists had discovered and charted the planet's most imposing landform—the Mid-Ocean Ridge, a globe-encircling chain of underwater peaks larger than the Rockies, Andes, and Himalayas combined. Experts also determined that narrow rift valleys run along the entire crest of this 46,000-mile chain. In the 1960s, a revelation occurred: The ocean floor was found to be expanding. It was moving outward in both directions from the rifts, driven by unknown forces.

"Drilling in the ocean basins revealed that new ocean floor is being formed," the Deckers explain. "These seafloor plates spread apart a few centimeters every year, and the growing gash is sealed from within by submarine lava flows." Here was a discovery as far-reaching as any in earth science, a mechanism that could account for the movement of continents, the distribution of earthquakes and volcanoes, and the formation of massive mountain systems.

The discovery of seafloor spreading at last gave weight to Wegener's idea of continental drift. "The new twist," the Deckers add, "is that the continents are locked into plates of spreading seafloor and move with, rather than through, the ocean crust." Those movements generate titanic forces. "Almost all the action is on the margins of the plates," explain the Deckers, "where they pull apart, shove together, or grind past one another."

On the edges of the Pacific Ocean, spreading seafloors collide with continental plates. There the ocean plates bend downward, or subduct, sliding beneath the continents. Pressures from such collisions warp and buckle the edges of the plates. The diving ocean crust forms offshore trenches, such as the western Pacific's Mariana Trench, an abyss so great that the conqueror of its 36,000-foot depths, the bathyscaph *Trieste,* needed nearly five hours to reach bottom.

As it overrides the ocean plates, continental crust is heaved upward, wrinkling into accordion-like folds of mountains such as the Andes. Peaks like those of the Appalachians form when one landmass smashes into another. Earth's mightiest range, the Himalayas, continues to be uplifted by the force of India pushing against the Eurasian continental plate. Most ranges result from such folding, but earth's shifting crust also thrusts up fault-block mountains: the Tetons, the Sierra Nevada. These form as blocks of crust are forced upward along fault lines.

Around the Pacific rim, friction between abutting plates triggers the earthquakes that plague areas such as Chile, Alaska, and Japan. As subducting plates plunge downward, they crack the earth's surface, allowing magma to escape. This tumult produced the Pacific's Ring of Fire, the volcano-dotted realm where more than half the world's earthquakes occur. The zone's unstable nature is a constant threat. In 1980, Washington's Mount St. Helens thundered to life in the Cascade Range. In mid-1985, earthquakes devastated parts of Mexico City. Later the same year, the eruption of Nevado del Ruiz, an Andean volcano, buried thousands of Colombian villagers under an avalanche of mud.

Curiously, the volcanoes of Hawaii do not lie along jostling crustal margins. Instead, they are found near the center of the huge Pacific plate. Yet their position offers persuasive evidence in support of plate-tectonic theory. That stationary "hot spot" described by Thomas Wright holds the key.

The man who first recognized the significance of hot spots is Canadian geophysicist J. Tuzo Wilson, retired director general of the Ontario Science Center. Dr. Wilson maintains that heat generated inside the earth must somehow escape, and that hot spots provide the outlet. "When the heat reaches the surface," says Dr. Wilson, "it cooks up lava and volcanic eruptions. Worldwide more than a hundred hot spots have been recognized, including the Yellowstone area in the United States and several in East Africa.

"It occurred to me that such a stationary hot spot could have created the Hawaiian Islands. In this hypothesis the plume pushes up island-forming lavas. In time the Pacific plate moves off, carrying the island along with it. Eventually the hot spot sends up another island."

Dr. Wilson points out that the islands and seamounts stretching away from the hot spot currently below Mauna Loa and Kilauea are progressively older, as they would be if formed by a single, fixed source. The most distant part of the chain is seventy million years old, while the island of Hawaii is around seven hundred thousand years old. Could a new island someday rise from the Pacific as the Big Island drifts beyond the hot spot? Scientists here think it possible, for even now underwater lava flows are building a seamount named Loihi some 19 miles southeast of Hawaii.

Besides their volcanic heritage, the Hawaiian Islands possess an ocean environment that makes them an ideal laboratory for varied investigations. Marine biologist Sylvia Earle has spent her life studying the sea. "To some, the abundance and, perhaps more significantly, the diversity of aquatic life are surprising," she says. "I am consistently amazed at the individuality of each member of each species. Just as we know that no two people are ever exactly alike, not even identical twins, we know that this is also true of . . . whales, fishes, and even hermit crabs."

Of all the life-forms in the oceans, few intrigue humans more than the mysterious "singing" humpback whale. Dan McSweeney has followed the eerie song of the humpback for more than 11 years. For the past eight years, the endangered creatures have lured the independent photographer-researcher back and forth between Hawaii and Alaska, end points in the migratory routes of a large number of the whales. "Humpbacks are found in all oceans," says Dan, "but the whales are thought to maintain distinct breeding populations." An estimated 10,000 humpbacks now exist, with several hundred returning to the Hawaiian Islands each winter to calve and possibly to mate. *(Continued on page 28)*

M onarch of the sea, a humpback whale surges to the surface in Alaska's Glacier Bay. The lower jaw of the 40-ton whale bulges with krill-laden water. Fringes of baleen suspended from the upper jaw strain krill and herring, the humpback's main diet. During summer feeding season, a humpback daily consumes more than a thousand pounds of food, storing fat that will sustain it through winter, the breeding season. The largest whales surpass all other animals on earth in size. The diverse life of the sea ranges from the ponderous to the ethereal. Delicate membranous "wings" propel Corolla spectabilis (opposite, left), an oceanic snail about four inches wide. No more than two inches across, a jellyfish—Orchestoma pileus (opposite, right)—swims by flexing its transparent dome.

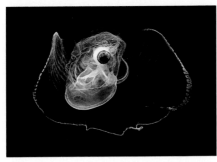

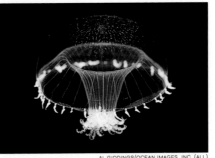

AL GIDDINGS/OCEAN IMAGES, INC. (ALL)

Hitching a ride on an iceberg, chin-strap penguins rest between feeding forays in Bransfield Strait, near the Antarctic Peninsula. Though Antarctica itself nurtures little life, mostly lichens and insects, its frigid, oxygen-rich waters support seals, fish, squid, and baleen whales. All these creatures, along with penguins and other seabirds, feed heavily on the Antarctic's enormous masses of krill. Euphausia superba, *two-inch-long krill (below, right), dine on algae—a food source found in winter on the underside of Antarctic ice.*

22

*P*rotected by a two-inch-wide carapace, a rock crab off Japan blends with an undersea ledge fringed with tentacles of a large sea anemone (above). Opposite, lower: Gaily striped clownfish in the Red Sea find refuge amid poisonous tentacles of a sea anemone. Scientists think the fish may build immunity to the anemone's sting by rubbing ever-larger portions of their bodies against the tentacles. Searching for hydroids and other soft-bodied invertebrates to eat, a rare sea spider (opposite, upper) scuttles along the bottom of Antarctic seas on 8-inch-long legs.

In Spooky Lake, one of 80 marine lakes in the Micronesian archipelago of Belau, a diver drifts near mangrove roots encrusted with mussels. The clear, brackish water fades into a layer clouded by algae. Several feet deeper, chemical reactions lace the depths with methane, hydrogen sulfide, and ammonia— gases all highly toxic to humans and other oxygen-breathing organisms. Below: A galaxy of harmless jellyfish, Mastigias, *engulfs the diver. These creatures, which inhabit three of Belau's lakes, perplex scientists.* Mastigias *mysteriously rotate counterclockwise as they swim. In each of the three lakes where* Mastigias *occur, they travel in different compass directions as they migrate from shore to opposite shore twice daily.*

FOLLOWING PAGES: Thrust above the sea by tectonic forces some 20 million years ago, the limestone isles of Belau wear a thick forest covering. Rain and rotting vegetation produce humic acids that eat through the limestone, cutting fissures and tunnels that allow an exchange of water between the lakes and the ocean. Each with its own physical, chemical, and biological properties, the saltwater lakes offer a natural laboratory for the study of oceanic systems in microcosm.

DAVID DOUBILET (BOTH); DOUGLAS FAULKNER (26-27)

Earlier, I had listened to recordings of humpback songs, complex sequences of sounds that can last half an hour or more. The humpback is one of the few whales known to sing. Researchers think this behavior indicates a high level of intelligence, for the phrases that make up a song are always repeated in the same order. I was struck by the beauty of these underwater arias, and by their amazing variety. A single song may shift from the haunting wail of an Indian sitar, to the playful booming of a jug band, to the chaotic screechings of a string section tuning up. I ask Dan the obvious question: Why do the whales sing?

"No one really knows," he says. "One theory is that, like birds, males attempt to attract females with their songs. A friend of mine, Canadian researcher Dr. Jim Darling, thinks the songs may be a way the males work out their dominance hierarchy, similar to the way mountain sheep establish their dominance through visual displays of their horns." Dan mentions that male humpbacks compete vigorously. "I've picked up some really awesome fights over the hydrophone. You hear these tremendous impacts between two whales—pow!—80 tons colliding. I've seen males surface with bloody dorsal fins and head knobs. This idea of the gentle giants isn't all that true."

Dan studies humpbacks from his 20-foot inflatable, *Black Whale.* He invites me onto the boat, showing me how to run it in case he gets into the water to take photographs. In addition to underwater pictures, Dan takes ID shots of the whales as they surface to breathe, concentrating on the dorsal fins and the undersides of the flukes. "Often the dorsal fins will have distinctive shapes, such as notches from old wounds. The flukes have black and white markings unique to each individual, sort of like fingerprints."

The first day on the water with Dan, I see no whales. It is April, late in the winter season, and many humpbacks have begun their migration to summer feeding grounds off Alaska. Our second day out, Dan spots an animal breaching in the distance. He turns the boat sharply, and we bounce across the choppy water. "That was a big one," he says. "Either a humpback or a sperm whale." When we reach the area, Dan drops the hydrophone over the side. Suddenly the speaker is alive with clicks like those of a Geiger counter. "Sperm whale," says Dan, explaining that only the echolocating toothed whales create such clicking sounds; baleen whales such as humpbacks do not.

Unfortunately, the sperm whale eludes us. But next day, our patrolling is rewarded. Late in the morning we spot a humpback and calf about a hundred yards offshore. Slowly maneuvering the *Black Whale,* Dan guides us to within a hundred feet. For the next hour, we observe these wondrous creatures in silence. The 40-foot adult swims protectively near her 16-foot calf. She surfaces every 15 minutes or so, while the calf comes up about every five. Breaking the surface, the whales spout with a loud sigh, taking three or four breaths before they dive. Spellbound, I watch the slow, majestic, Ferris-wheel roll of the adult. First the lumpy head breaks the surface, followed by the broad, shiny back that seems to slide on forever. Finally, with the signature flip of her mighty flukes, she slips below.

As we head back to port, Dan comments, "It's really amazing when you think about it. Here we are, specks on the planet, trying to come together with these other specks, creatures that spend only 5 percent of their lives near the surface. The frustrating thing is that there's just this fragment of time in which you're actually witness to anything. But it's fantastic to see something for whatever moment in an environment like the ocean, which man still hasn't been able to penetrate to a large degree."

Sylvia Earle concurs. "The liquid realm around us remains nearly as unprobed as the distant planets," she says. It is ironic that earth's most significant feature should remain its least known. Three-quarters of the world's surface is covered by water or mantled in ice, yet we are still in the infancy of our knowledge of these areas. For example, we continue to discover more about how the oceans and polar realms influence the planet's weather and longer-term climate.

The relationship of the oceans to weather and climate is a prime topic of study at the Scripps Institute of Oceanography in La Jolla, California. At Scripps, I meet the father of long-range weather forecasting, Dr. Jerome Namias. He began studying weather as a high school student in Fall River, Massachusetts. Some 60 years later, he is still learning about what he calls "the second most complex system on earth—after human behavior."

I ask Dr. Namias, the founding chief of the U.S. Weather Bureau's extended forecast division, to sketch the forces that power earth's weather machine. "The sun's the instigator of it all," he says. "The sun heats the tropics, while the poles lose heat, in part by reflection. This creates an imbalance that has to be made up. Otherwise the tropics would get hotter and hotter and the poles would get colder and colder. Nature takes care of this imbalance by forming eddies in the atmosphere, a large-scale turbulence that mixes everything up."

Dr. Namias explains that the mixing occurs as warm tropical air rises, forms clouds laden with moisture, and moves toward the poles; cold air—drier and heavier—moves in to take its place. At the poles, the same cycle occurs: Cold air moving toward the Equator is replaced by warmer air. This exchange of heat helps create earth's prevailing winds: the polar easterlies, prevailing westerlies, and trade winds. It also helps power global water circulation: Water evaporation from the ocean drifts inland as fog or clouds. Where air masses with different temperatures collide, storms and precipitation result.

Understanding earth's general air circulation remains "the central problem of all meteorology," says Dr. Namias. Despite modern instruments, reliably predicting the movements of shifting air masses remains a difficult task. "We're dealing with an atmosphere that is a fluid, one that has hydrodynamic chaos in it."

In a planetary cipher, recurring patterns of air circulation scribe their movements on the land below. As it rises, the warm, moist air of the tropics dumps drenching rains that nurture lush forests such as those of the Amazon basin and equatorial Africa. Emptied of its bounty of life-giving moisture, the dry air sinks as it continues poleward, creating a band of parched landscapes—the Arabian Desert and the Sahara, for example. Touching every continent, the deserts of the world lock a third of the land in their withering grip.

Beyond the zones of aridity lie the fertile lands of the temperate regions. Here extremes of moisture or dryness are lessened by the prevailing westerlies, which receive infusions of both tropical and polar air. In this milder climate flourish temperate forests and grasslands, grasses dominating in areas where moisture is insufficient to sustain trees.

Tropical and temperate forests together green an area more than five times the size of the U.S. The world's forests support an amazing range of wildlife, from the untold thousands of species that dwell within tropical rain forests to the brief inventory of creatures that inhabit the boreal forests of northern latitudes.

Grasslands, among earth's most fertile areas, have long been critical to man's survival, yielding the grains that furnish the bread of life. Grasslands also sustain tremendous numbers of animals, both wild and domesticated. The herds of bison that once thundered across the North American prairie are legendary. On Africa's vast Serengeti Plain, masses of wildebeest still darken the land as they migrate in the wake of seasonal rains.

Pondering the forces that shape the planet's weather patterns, Dr. Namias came to the conclusion in the late 1950s that certain "boundary influences" play a critical role in weather. Those influences include the seas and snow or ice cover. They affect weather by exchanging heat and moisture with the atmosphere. "It became apparent that the ocean contained the potential for affecting the weather substantially over vast areas," Dr. Namias says, noting that the northern Pacific influences much of North America's weather.

He points out that the sea changes temperature slowly and can store massive amounts of heat over large areas for many months. "Just the top three meters of the oceans alone contain as much heat as the entire overlying column of air." This stored heat has a stabilizing influence on the world's weather, moderating the climate in many parts of the globe.

The ocean transports heat in sweeping currents such as the Gulf Stream, whose warming waters help temper the climate of Europe. Mimicking atmospheric patterns, cold polar water moves toward the Equator at great depths. In places, currents of cold water well upward, as along the Pacific coast of South America. There the cold water cools the atmosphere, inhibiting rain. The result is an arid coastal strip, the Atacama and Peruvian Deserts. Other deserts, such as parts of the Great Basin in the American Southwest, form in the leeward rain shadow of mountains, which trap moisture from inland-drifting oceanic air.

At times, an area's normal climate mysteriously alters. One of the most puzzling aberrant weather patterns is called El Niño. Occurring irregularly, El Niño often produces devastating effects along the Pacific coast of South America. The waters off Peru—among the world's richest fishing grounds—suddenly become barren, and torrential downpours cause flooding along that country's arid coast. This freakish pattern may last for months and can trigger weather shifts thousands of miles away. Just what causes El Niño is still the subject of intense study.

Scientists investigating climatic behavior often turn to the past. Clues can turn up in unlikely places. Growth rings of the long-lived bristlecone pine, for instance, provide a yearly record of temperature and precipitation. Drought periods produce narrower rings than those of wetter years. Core samples drilled in the Antarctic and Greenland ice sheets also record past conditions, including catastrophes such as volcanic eruptions. Debris from an eruption settling on snow cover may show up as a dark line in an ice core taken thousands of years later.

By patiently analyzing such evidence, researchers have pieced together an outline of earlier climates. They have found that the earth is cooler today than in previous periods. During the past billion years, the world's average temperature is estimated to have been about 72°F much of the time; today's global average is 58°F. In those warmer periods the poles were ice free. Occasionally, for unknown reasons, the warmth was interrupted by shorter periods of cooling. Then massive sheets of ice crept outward from the poles, covering large portions of the earth's surface. The glaciers formed as snowfall accumulated year after year, gradually being compressed into ice crystals under its own immense weight.

Climatologists believe at least four major periods of glaciation have taken place in the past billion years. The most recent, the Pleistocene epoch, began some two million years ago. At its maximum, the Pleistocene ice sheet dipped deep into North America, reaching as far south as present-day Missouri. As it advanced, the ice sheet reshaped the land, carving valleys, grinding down mountains. It also forced many plant and animal species to migrate southward. At one time, boreal forest and tundra covered the southern Appalachians.

The final mantle of Pleistocene ice began its retreat some 18,000 years ago, as global temperatures grew warmer. But even today, the earth retains vestiges of that ancient ice. "The oldest ice core that's datable is that being obtained now by the Russians at Vostok research station, in Antarctica. They're more than a hundred thousand years down, and they're still going."

The speaker is Mark F. Meier, director of the Institute for Arctic and Alpine Research at the University of Colorado. "Pleistocene ice has been sampled and studied in Greenland as well as in Antarctica," he continues. "We know its characteristics. It's different from present-day ice. It's dirtier; its grain sizes are smaller; it's softer and flows more easily."

Although the great ice sheets have retreated, global temperatures are cool enough to sustain enormous amounts of ice. In 1960, earth's coldest recorded temperature, -127°F, numbed Vostok station. Ice as much as two miles thick covers Antarctica and Greenland, and an expanse of sea ice blankets the Arctic Ocean most of the year. So great is the weight of ice on Antarctica that a third of its land surface sags more than a thousand feet below sea level.

Dr. Meier points out that while glaciers shrink from a combination of warmer temperatures and decreased precipitation, a global cooling trend and increased precipitation could trigger a dramatic growth in glacial ice. If the amount of solar energy reaching the earth decreased a mere 2 percent, some experts say, the oceans would freeze and snow would lie on the Equator.

Today, some 10 percent of the planet's land lies beneath ice. Antarctica, by far, accounts for most of that expanse. Mountain glaciers also grip large areas. Unlike ice sheets, which inch outward in all directions, mountain glaciers are confined within a path that directs them. Grinding through channels of stone, the moving ice helps produce such breathtaking scenery as California's Yosemite Valley and the alpine wonders of Switzerland.

For Mark Meier and other glaciologists, the ice-scoured mountains of Alaska represent a living laboratory. Many researchers believe that Alaska may be a crazy quilt of 50 terranes, remnants of ancient landforms shoved together during the past 150 million years by earth's shifting crust. Collisions among those drifting chunks of real estate helped forge today's tortuous landscape, from the Brooks Range in the north to the St. Elias Mountains in the south.

In Alaska, scientists can observe ice-age forces at work, for the retreat or advance of glaciers there often takes place on a human *(Continued on page 38)*

With a thunderous roar, an iceberg tumbles from Margerie Glacier in Alaska's Glacier Bay. Two centuries ago, a 4,000-foot-thick tongue of ice clogged all of Glacier Bay. The ice has since retreated 60 miles. Such rapid wasting results when melting and calving exceed the rate of snow accumulation. Below, a hiker marvels at a translucent ice cave near Muir Glacier, named for the naturalist who—in 1879—first mapped Glacier Bay.

FOLLOWING PAGES: Jakobshavns Glacier juts into Disko Bay in Greenland. Glaciers and polar ice bounce the sun's rays back into space. Earth's remaining surface area absorbs heat unevenly; this imbalance sets in motion the air and ocean currents.

*A*s if at earth's end, a
youngster peers over the
brink of Prestestolen
Rock, a sheer granite
outcrop 2,000 feet above
the waters of Lysefjord on
the west coast of Norway.
Ice Age glaciers scoured
much of northern
Europe. The thickest ice
flow—a mile and a half
deep—covered present-
day Norway. There
glaciers carved deep
fjords out of solid rock.
When the ice retreated,
seawater filled these
steep-sided clefts to depths
of more than a thousand
feet. Among earth's most
powerful forces, moving
ice has molded most of
North America's face, as
far south as the Ohio and
Missouri River Valleys.

scale rather than in the slow sweep of geologic time. In 1984 alone, Alaska's mighty Columbia Glacier retreated nearly three-quarters of a mile. The year before, the "surging" Variegated Glacier advanced more than 200 feet a day, a hundred times the average speed that glaciers travel.

Dr. Meier estimates that 75 percent of the world's fresh water is locked in glacial ice. Melting glaciers free tremendous amounts of water. Average sea level has risen 4 to 6 inches over the past century, an increase Dr. Meier attributes partly to melting mountain glaciers. "An incredible 40 percent of the water returned to the oceans by glaciers comes from the mountains edging the Gulf of Alaska in British Columbia and Alaska. That region contains huge masses of ice." The Bering Glacier alone covers an area as big as Delaware. "Those glaciers have been wasting rapidly, while the glaciers in the polar regions haven't changed much in a hundred years," since weather conditions there are more stable.

Along the southern panhandle of Alaska, I visit an icy wonderland, starkly beautiful Glacier Bay. Within this deep slash in Alaska's rugged coastline, tidewater glaciers rumble down from the snowcapped Fairweather Range to calve enormous icebergs into the sea. John Muir visited this region often, first coming here in 1879. He described Glacier Bay as "a solitude of ice and snow and newborn rocks, dim, dreary, mysterious." Protected now as a national park and preserve, Glacier Bay still exudes an aura of mystery in its enveloping silence.

At Bartlett Cove, near the bay's mouth, I camp in a dense spruce-hemlock forest. Only occasionally do I hear the cries of birds, eerie sounds in the chill, misty air. From the woods comes the mournful hooting of the blue grouse; along the shore, the cockney slang of strutting ravens; and over the bay, the argumentative babble of gulls, pierced by the thin, distant screams of circling eagles.

M y tent stands on historic ground at Bartlett Cove, for here is the terminal moraine of the 20-mile-wide tongue of ice that once filled all of Glacier Bay. The moraine, the rocky payload carried and pushed by the advancing ice, was left behind when the glacier began a retreat up the bay some 200 years ago. The barren rubble was eventually colonized by algae and mosses, followed in slow succession by scouring rush and fireweed, then by dryas, alder, willows, spruce, and hemlock.

I walk now through a maritime rain forest, a tangle of arrow-straight giants and fallen logs. Everywhere I see wisps and clumps of moss—clinging to tree limbs like limp green rags hung out to dry and covering the ground in a cushioning layer. Along the shoreline, stands of trees grow near the water's edge. Those closest to the water are overshadowed by successively higher ranks. In reality, the trees are the same height and age: The inland trees are being raised above those nearer the shore. Relieved of the tremendous weight of ice it formerly bore, the land here is rising an astounding inch and a half a year.

When John Muir first visited Glacier Bay, the ice had retreated about 40 miles from the Bartlett Cove area. A century later, measurements show that it has receded some 20 miles more—in all, the world's greatest recorded glacial retreat. Boarding a boat, I journey up-bay to view a sampling of the park's 16 actively calving glaciers. Along the way, I am witness to an unusual phenomenon: plant succession in reverse. Ever-sparser plant coverage denotes areas more recently freed of ice. The forest of Bartlett Cove gives way to alder thickets, which in turn are replaced by dryas and fireweed. Farther up-bay, mosses scrabble tenaciously for a hold. And then, all is bare rock—and ice.

As we motor past black bears lumbering along the shore and mountain goats clattering across rocky cliffsides, I recall descriptions of glaciers I have read. Muir talked of vast ice walls in "many shades of blue, from pale, shimmering, limpid tones . . . to the most startling, chilling, almost shrieking vitriol blue." Yet nothing prepares me for the majesty of Muir Glacier, the 15-story-high river of ice named in honor of the pioneering naturalist.

Entering the fjord of Muir Glacier, our boat eases through the berg-choked water. Chunks of ice bounce off our steel hull with an ominous clunk. We stop a quarter mile from the ice front. There the engine is turned off, and we rock in silence as a chorus of glacial hosannas echoes down the rocky gorge—hisses, groans, and resounding thunderclaps produced by the stresses of moving ice.

The electric blue surface of the glacier is deeply fissured, broken by crevasses that form as the ice heaves its way over uneven terrain on its long downhill march. The unreal color results from the density of the ice, which absorbs all colors but blue. Cloudlike, the jumbled surface takes on fanciful shapes: a forest of azure sequoias, a crumbling castle of blue.

Undermined by water at the glacier's terminus, towering bergs calve with the crack of cannon fire. Twice, columns of ice the height of the entire glacier fall away, one accompanied by the sudden explosion of an "upper"—a berg that calves below the water and rockets to the surface. Each time a slab splits from the glacier, I think of the centuries it took to build. Silently deepening, each season's snows were like the efforts of those medieval laborers who toiled to build an edifice they would never see completed. Now, with a heaven-rending roar, the patient handiwork of untold years drops from the glacier's face, an icy Notre Dame sliding into the sea.

Amid the fjords and mountains of Alaska, the earth-shaping power of ice is inescapable. Yet I recall evidence of glacial sculpturing in less dramatic settings. It is seen in New York City's Central Park, where boulders were plunked down by a retreating glacier; in Minnesota and the Dakotas, where kettle lakes were left by melting remnants of ice; and throughout the Midwest, where fertile glacial sediment has created a nation's breadbasket.

However powerful, glaciers are but one tool nature uses to reshape the land. Wind and water serve equally well. The frightening power of wind erosion was blasted into this nation's memory in the 1930s, when Dust Bowl storms carried away tons of topsoil, leaving parts of the Great Plains nearly as desolate as Africa's Sahel region. Water, too, can move mountains of soil. One 1985 study estimated that, each year, more than six billion tons of soil are lost through water erosion in the United States alone.

Few places reveal the erosive force of water better than Arizona's Grand Canyon. To stand on the rim of the Grand Canyon is to gain perspective on the planet's 4.6-billion-year history. Civilization's 10,000-year run seems a trifling entry as you stare into this great dusty volume of time. Rocks nearly two billion years old have been exposed in the canyon's depths. Averaging 10 miles wide, this 277-mile-long chasm is still being cut by the grating, silt-laden Colorado River, by abrasive grit borne on the wind, by the rock-shattering power of freezing water.

It is cool now as I stand on the South Rim of the canyon. The morning sun has not yet cleared the horizon, and spread before me is a shadowy world of untold dimensions. Today, I will hike to the bottom of the canyon—a journey down Bright Angel Trail, which zigzags for eight miles to the edge of the Colorado some 5,000 feet below. By the time I reach the river, the temperature will have soared to nearly a hundred degrees. I carry a full canteen and wear a shade-giving hat. Yet before I have gone far I am perspiring, and I lift my canteen more and more frequently as the rising sun warms the dry earth beneath my feet.

Hiking along, I gaze about in fascination. A fairyland carved and painted with a master's touch emerges in the soft morning light. Battlements and pinnacles of stone have been shaped with effortless artistry, then colored with earth-hued reds, oranges, purples, and golds. The variety of shapes and colors seems chaotic, but the rocks teach a mute lesson in the physics of erosion: Steep-sided cliffs are made of resistant limestones and sandstones; gentler slopes, covered with rocky rubble, are formed of more easily eroded shale. Narrowing gradually, the canyon at last reaches the inner gorge of the Colorado. There, deep-lying metamorphic rocks resist the cutting power of moving water, forming the sheer walls familiar to river runners.

After four hours of hiking, I enter the inner gorge and see the tumbling brown waters of the Colorado below. Hot and tired, I switchback down the trail, then cross a footbridge over the river. Plopping down in a patch of shade, I kick off my shoes and stare at the river's roiling surface. I listen long to the mighty river as it courses on its way from the distant Rockies to the Gulf of California. Suddenly it occurs to me that I am observing the completion of a cycle, a loop of forces interacting with miraculous precision.

Through my travels, I have learned of the earth-wrenching power of plate tectonics—how spreading seafloors shift the continents, trigger earthquakes and volcanoes, and build mighty mountain ranges. I have seen that the sea yields up its moisture to the land, precipitation that nurtures life in forests, in grasslands, even in forbidding desert realms. As it moves, water also reshapes the land, carving canyons and wearing down mountains. Now, gazing at the Colorado, I am witnessing the journey of that water homeward toward the sea. And as it surges past, it carries with it a freight of sediments that will create new landscapes. Building, destroying . . . building yet again—it is all part of the mystery and splendor of a world perpetually in the making.

Hikers in Utah explore Buckskin Gulch, a streambed in the Paria River drainage area. These sandstone walls, smoothed by centuries of summer torrents, record the power of moving water. Flowing to join the Colorado River near the Grand Canyon, the river and its tributaries helped carve Utah's canyonlands.

PAGES 42-43: Lightning dances above the North Rim of the Grand Canyon. A mile below curls the Colorado River, progenitor of this fantasyland of stone—the world's most spectacular example of erosion.

TOM BEAN (OPPOSITE AND 42-43)

*L*and of Room Enough and Time Enough—so the Navajo named Monument Valley, an otherworldly realm on the Utah-Arizona border. Some 100 million years ago, the land here stood level with the tops of these thousand-foot buttes, but wind and water carried away sands and shales, leaving monoliths of harder sandstone.

PAGES 46-47: Without safety rope, world-class climber Ron Kauk free-solos at 2,000 feet alongside California's Yosemite Falls. Beyond, the weathering granite face of Half Dome testifies to the earth's slow but ceaseless change.

Lofty Titans

The Mountains

By Cynthia Russ Ramsay

Framed by Douglas firs and flowering nuttalls, Mount St. Helens slumbers in Washington's Cascade Range— two years before its cataclysmic eruption. Beneath the oceans, volcanic mountains push isles of paradise above the surface. On the continents, volcanoes and other peaks lift fascinating worlds above the clouds.

FOLLOWING PAGES: After the eruption, one of the largest in U.S. history, a gaping crater replaces Mount St. Helens' snow-capped peak. On May 18, 1980, the volcano exploded and turned a picture-postcard landscape into an ashen moonscape.

DAVID MUENCH (LEFT); GARY BRAASCH (50-51)

louds of volcanic ash and gas drift past a scientist in a lifeless grove of aspens and poplars on Mount Usu in Japan. Exploding in 1977 and 1978, Usu unleashed mud slides that destroyed the town of Poyako-Onsel. On the eastern flank of Mount Usu, volcanologists (opposite) set up instruments to analyze gases escaping from a vent, the area's hottest at 752°F. By determining the composition and temperature of hot gases from molten rock deep within the earth, scientists may gain advance warnings of eruptions. Hundreds of active volcanoes in the mountain systems bordering the Pacific Ocean form the Ring of Fire.

The Mountains

They chose to die, awaiting death as the tinkling music of the gamelan orchestra sounded through the village. Clad in their best sarongs, they brought offerings of rice and flowers to the gods. Some carried lanterns, for the black cloud of smoke and ash had turned day into dusk. Molten rock spewed out of Mount Agung's crater and surged down the slopes in fiery tentacles, suffusing the darkness with a lurid glow. Trees ignited at the lava's touch. Streams hissed into vapor, and the air reeked with the stench of sulfur.

Led by their village headman, scores of people in Badeg Dukuh, a village high on the flanks of Bali's sacred Mount Agung, refused to flee the inferno; they were reluctant to offend the gods by abandoning the 10,308-foot volcano, where their deities dwell. A few hours after the beginning of a devastating eruption on March 17, 1963, searing clouds of gas and ash engulfed the village, silencing the cymbals, drums, and xylophones and snuffing out the lives of those who stayed.

While many people fled the danger, there were many more in other upland villages who stood fast. Some were too frightened to leave; others simply refused to take refuge from the wrath of their gods. They also met their doom, suffocated by the hot, gaseous clouds known as *nueés ardentes.*

I learned of these tragedies from my guide, Made Yastina, as we traveled around palm-fringed Bali—one of the thousands of islands in the Indonesian archipelago, which straddles the Equator for more than 3,000 miles. I had come to this region of 128 volcanoes (one of the most active volcanic zones on earth) in a far-flung investigation of the mountain world. Elsewhere, in the lofty reaches of the Himalayas, Andes, and Rocky Mountains, in the wild, windblown realm above tree line, I discovered plants and animals that have adapted to the harsh conditions of life in remarkable ways.

From Mount Kinabalu on the island of Borneo, with its eerie landscape and puzzling flora, to the jagged summits of Wyoming's Tetons, I came to know the heights as places of challenge and adventure, as places that stir the imagination with stunning scenery. Most of all, my forays into the mountains taught me that they are places where nature reigns unchallenged.

Nowhere is nature's awesome power more evident than with volcanoes. Twenty-two years after Agung exploded, some of the destruction was still visible. Bands of dark lava blighted the upper slopes. Boulders that had been catapulted from the crater bore witness to the volcano's fury. Near the town of Klungkung, rice fields lay buried beneath a thick layer of volcanic ash. Villagers had been mining the deposits since 1972, and we paused to watch workers fill up dump truck after dump truck in the huge, dusty pit. "The volcanic sands and gravel make good concrete," Made explained. "The people are earning more money this way than they would by farming."

Elsewhere on the island the eruption had enriched the fields in a more traditional way—fertilizing the soil with minerals from deep within the volcano's fiery heart. In most places the desolate landscape had returned to lush productivity. Wide, curving rice fields in various stages of growth terraced the lower slopes with stairways of ripe, golden grain, tiers of young, jade-green plants, and row after row of flooded paddies, their glossy surfaces reflecting the sky and the passage of scudding clouds.

Nearby, on the densely populated island of Java, 16 active volcanoes have created some of the best farmland on earth. More than 200,000 people risk living around smoking Merapi—the Fire Mountain. The 9,551-foot peak has killed many times since the 1800s. Almost every year the village *tong-tongs* sound their alarm. When the log drums boom through the heavy tropic heat, people flee their homes, but they return as soon as they can.

"The long-term blessings count more than the occasional misfortune," said Nana Rahnama. He was accompanying me from the city of Yogyakarta to the isolated village of Kinarejo, which sits five miles from the sulfurous crater. Ours was a surprise visit to Maridjan, an elder who has been entrusted with the honor and responsibility of bearing annual ritual offerings to the crater for the gods of Merapi. He acts on behalf of the venerated sultan of Yogyakarta—the ceremonial leader of millions. Inside Maridjan's home, cool and airy with its rattan walls, tall bamboo beams, and high-pitched tile roof, the elder spoke of signs that alert him to the possibility of an eruption: the way the corn grows, the way birds fly, and how snakes behave.

Is it all superstition? Director of the Volcanological Survey of Indonesia, Dr. Adjat Sudradjat, an urbane and respected scientist, has from time to time wondered if, for example, snakes couldn't feel subtle temperature changes in the earth that send them slithering away before an eruption.

Dr. Sudradjat, however, relies on a staff of scientists and technologists who use sensitive instruments to monitor Merapi, one of Indonesia's most menacing volcanoes. The danger signals are increased seismic activity, an increase in the temperature of gases in the crater, or a bulge or change in a mountain's slope. But volcanologists are quick to admit there are limits to their readings.

"We can detect departures from the normal, but we still don't know precisely what makes a volcano go off when it does," volcanologist Robert I. Tilling of the U.S. Geological Survey had told me.

Most active volcanoes lie along the boundaries where tectonic plates collide. The crash of these huge slabs of the earth's crust forces one of the plates to sink, or subduct, below the other. The titanic pressures of the subduction melt the rock, and it begins rising intermittently until it eventually explodes through the earth's surface.

Indonesia has about 70 of the world's more than 500 active volcanoes. Why Indonesia contains so many has baffled earth scientists for some time; one explanation may be that the plates are moving more rapidly at this junction. "The plate that underlies the Indian Ocean floor south of this region is both old and dense. Because of its density, it subducts at a high rate under the Indonesian island arc," says geologist Haraldur Sigurdsson at the University of Rhode Island.

Paradoxically, the pall of ash and the rivers of molten rock freed by volcanic explosions are indispensable to life on the planet. Were it not for volcanoes, scientists theorize, there would be no water, no atmosphere. Life as we know it would not exist. Geologist Sigurdsson explained why. "During the early years of the planet—it's around 4.6 billion years old now—there was no atmosphere. Then volcanic vents began releasing gases, mainly carbon dioxide and steam pent up in the earth's interior." The volume of water liberated as steam in an eruption is astonishing. For example, Parícutin, a volcano that in a matter of weeks sprouted out of a Mexican cornfield in 1943, gushed great volumes of water vapor. "Steam from these primeval volcanoes," Haraldur continued, "eventually condensed into oceans. But the earth had no free oxygen until it was liberated by blue-green algae. These organisms took the carbon dioxide in the primitive atmosphere and used energy from sunlight in manufacturing sugar and starches. That miracle of photosynthesis released oxygen, and in about two billion years enough had accumulated to produce the atmosphere we have today."

Volcanism has occurred virtually everywhere. Dig a deep hole in Kansas or almost anyplace else, and you hit volcanic rock, for about 80 percent of the earth's crust is of volcanic origin.

Volcanoes reach their greatest heights in South America's Andes. The summits of such giants as 20,702-foot Chimborazo and 19,347-foot Cotopaxi lift ice-and-snow-sheathed peaks above the steamy floor of the Amazon basin. Nearly 60 active volcanoes and many more inactive ones are strung along the 4,500-mile-length of the Andes. Here in South America, the crash of tectonic plates that melted rock and created volcanoes in the Andes also crumpled and buckled the earth's crust to form this mighty chain of folded mountains along the western edge of the continent. Frequent earthquakes indicate that the cordillera is still under construction.

In Venezuela a half-day excursion by cable car takes you from Mérida, a mile-high city surrounded by coffee plantations, to the glaciers and jagged summit of 16,427-foot Pico Bolívar. The first in a series of four trams swings out above the deep gorge of the Río Chama and moves smoothly upward past a misty cloud forest, where plants grow in almost macabre profusion. Leafy red bromeliads entangle the topmost branches; lichens cascade down lower limbs; orchid vines twist around tree trunks, and mosses smother the damp ground.

Higher up, as it becomes cooler, trees are shorter—their leaves smaller and darker. Gradually the pandemonium of growth thins out, and your tram breaks through the mist into brilliant sunshine. You are looking down at the open expanse of the paramo, a land where it is too cold for trees to grow. The cloud-swept world seems very far away in this stark landscape. No matter what the season, intense solar radiation turns every day into scorching summer, and each night brings bitter winter.

This is the domain of the *Espeletia*. These plants, oversize members of the daisy family, grow as tall as people. Locals call them *frailejones*—tall friars. To me they look like yellow-green artichokes atop a shaggy stem. Scientists see the strange forms as adaptations to the environment. Dr. Tom Givnish, a University of Wisconsin ecologist, points out that by getting off the ground and growing tall the plant escapes the extremes of air temperature at the surface. The surface air can heat up to well over 100°F during the day and can fall below freezing at night. The ground can actually be 50 degrees warmer than the air just three feet above it.

"In Africa and Hawaii unrelated or distantly related plants have developed strikingly similar tall forms to meet the same conditions," Tom said. "This strange phenomenon of small herbs becoming giant plants occurs only at high elevations on the Equator. The *Espeletia* protects itself in other ways. The emphasis is on insulation from the daily thermal extremes. Its rosette of sword-shaped leaves closes around its bud at night to keep it from freezing and opens again as soon as daylight comes. The dead leaves don't drop off. They stay and form a blanket that prevents the stem from freezing. Like many plants surviving on the alpine tundra, the *Espeletia* has fuzzy leaves; the downy hairs form a cocoon that deflects wind, traps heat, and reflects ultraviolet radiation, which is considerably greater here than at sea level."

Your tram ride of nearly eight miles ends in a zone where only lichens grow. Just ahead the naked fangs of Pico Bolívar graze the sky, and the summit is only a short but challenging scramble away.

Strange plants and animals set the Andes apart and add to their fascination. From mossy cloud forests and sunlit glaciers at the Equator to wind-raked pinnacles and stormy fjords of Tierra del Fuego, where penguins waddle and condors soar, the mountains harbor realms of astonishing diversity.

Nowhere else would I see such a succession of scenes as on the lofty puna. This plateau, colder, drier, and bleaker than the paramo, sprawls across parts of Bolivia, Peru, Chile, and Argentina. Hundreds of flamingos strut and preen their rosy-pink feathers in the blood-red waters of Bolivia's Laguna Colorada. Billions of algae give the lake its nightmarish hue. Bells tinkle in the cutting wind as a caravan of haughty llamas files across a pass just below snow line. Sometimes called "camels of the clouds," these surefooted beasts thrive where the air is thin. Related to the gazelle-like vicuñas, they have red blood cells that can capture more oxygen than those of most other mammals.

In isolated corners, amid the rubble of stone and clumps of yellow ichu grasses, columns of flowers tower 30 to 40 feet—a startling contrast to the puna's starkness. The stalk bearing thousands of white blossoms identifies the plant as the giant puya—a relative of the pineapple. The puya blooms infrequently, perhaps less than once a century. No one can say how these fantastic plants grow to such mammoth size with so little sustenance in the scant soil of the puna.

Many unanswered questions remain about how plants manage to live in the harsh conditions of high places. Consider the alpine buttercup in North America; the plant can bud under several feet of snow. "How does it know it is time to bloom?" Louisa Willcox, director of field studies at the Teton Science School near Jackson, Wyoming, poses the question. "The human eye cannot detect light through more than about a foot of snow. Either the buttercup has a very sensitive means of detecting light, or it reacts to infrared radiation, or something else," she said.

We were strolling through an alpine meadow flush with flowers. Across the undulating terrain, the land dissolved into a blur of rainbow hues. Close up we saw yellow cinquefoils, blue lupines and forget-me-nots, purple vetches and rosy pussytoes. Each tiny flower was a dab of intense color against the pale ground.

It was early July in the Tetons, and at 10,450 feet on Rendezvous Mountain patches of wet snow lingered in the lee of outcrops and on north-facing slopes. The snow glared from couloirs and deep recesses in *(Continued on page 63)*

*P*rayerful Indonesians (opposite) at crater's edge make offerings of money, flowers, food, even a bottle of orange soda (above). They hope their gifts will appease angry gods they believe dwell within Mount Bromo, one of 16 active volcanoes on Java, an island the size of Arkansas. The Indian Ocean archipelago of Indonesia counts 128 volcanoes, each an influence in the life of the islanders. Every village has a dukun, or soothsayer, to foretell eruptions by his dreams.

PAGES 60-61: Miners rest on chunks of sulfur in a crater on the side of an active volcano on Java. Large tubes direct the flow of molten sulfur out of fissures. The sulfur solidifies as it cools and tumbles out, kicking up powdery clouds. In mining this nonmetallic element—used to make such products as medicines, paper, and gunpowder—workers must fight off nausea caused by its rotten-egg stench. To lessen their distress, they go to work on an empty stomach.

the rough gray rock, highlighting the sheer lines of the bare summits. In the alpine tundra, which begins at about 9,500 feet in the Tetons—at 12,000 feet in the Sierra Nevada and at 4,000 feet in the Canadian Rockies—summers are short; snow can come any time of year. In the spring, snowmelt saturates the ground, but ordinarily the soil is too scant to hold moisture. So the plants must cope with both flood and drought. And year-round the wind rarely stops.

"The primary strategy up here is to get small and get out of the desiccating wind," said Louisa. She was pointing to a moss campion, a cushion of tiny green leaves studded with flowers like pink confetti. The cluster of plants was only two inches above the ground, but the roots were about two feet long—proportions suited to finding water and to hanging on despite storms and rock slides. Moss campion and other cushion plants grow in tight colonies, like animals huddling together for warmth. With surface temperatures well below freezing, it is as much as 20°F warmer inside the tangle of vegetation.

We entered Cody Basin, where runoff from snowmelt has created an alluring pool. The rocky ground here also has a story to tell. In the limestone debris we found fossil shells and coral from the times when the Tetons and all the Rockies lay beneath vast, shallow seas. The Rockies stabilized 50 million years ago, geologists say, and now most of them are wearing down as ice, wind, and water nibble and gnaw at the mountains. But the upheavals that lifted the Tetons to their magnificent heights began about ten million years ago and continue today.

Wyoming geologist J. David Love explained that most mountain ranges in the Rockies have been folded up into great arches. The Tetons, however, were punched up and tilted westward along a fault, a fracture in the earth's crust. "At the same time the floor of Jackson Hole was dropping like a giant hinged trapdoor. The rising mountains and sinking valley are still moving, producing a vertical displacement averaging one foot every 300 years," David said.

A 1983 discovery in Jenny Lake, however, hints of more violent activity. Several 70-foot skeletal trees have been found standing upright in the lake and rooted, some divers say, in the bottom muds. Only the calmest water betrays the presence of the spectral shapes, carbon dated at 600 years old.

"If the trees are actually rooted," David said, "it means they were growing on land that has dropped 80 feet in the last 600 years—slammed down by what must have been some monstrous earthquakes. If they are not rooted, they must have slid from the mountainsides into the lake and somehow landed upright."

We were looking west across the valley of Jackson Hole to the granite spires of the Tetons. With sudden steepness the mountains soar from a land of unusual flatness, and there are no intervening foothills to diminish the visual impact of

"Child of Krakatau," Anak Krakatau fumes between Java and Sumatra. This lava cone, now 650 feet high, began to grow from the caldera of Krakatau, a volcano that exploded and collapsed into the sea in 1883. The eruption claimed 36,000 lives in nature's mightiest upheaval of modern times. The explosion spawned giant sea waves that ravaged islands and hurled a steamship two miles inland.

the peaks. The snowcapped mountains were so beguiling that I almost over-looked the herd of elk foraging for grasses just yards from my feet.

Like cardboard cutouts the Tetons have little depth. From their summits you can look down at Idaho potato fields that stretch away to the west. East of the mountains, a highway runs along the 40-mile length of the range, and jets land close to the base of 13,770-foot Grand Teton.

In contrast, it takes a long day's hike or horseback ride to find the drama of Wyoming's Wind River Range. This wild region southeast of the Tetons encom-passes tier after tier of mountain ridges, dense forests, and clear lakes. Bighorn sheep gambol on nearly vertical cliffs. Grizzlies paw the ground for roots and ber-ries. Perhaps even Bigfoot prowls the deep canyon recesses of the range.

Does the elusive creature exist? Is it a hoax or a scientific mystery?

"There are some very specific accounts of sightings in the Winds and in re-mote regions of California and the Northwest—where the creature is known as Sasquatch. The sightings are remarkably consistent and show a pattern in keeping with zoological principles," said Dr. George Gill, a physical anthropologist at the University of Wyoming.

Dr. Gill has analyzed the reports and studied the data on the tracks. "If those tracks are genuine, as some anthropologists believe, only a two-legged apelike creature of giant size described in the accounts could have made them," he told me. "We have a perfect fossil ancestor for this form in *Gigantopithecus,* which lived in India and China about half a million years ago. What makes many scien-tists skeptical is that not a single bone has been found on this continent to provide solid evidence. But after all, we have only a few bits of jawbone and teeth from *Gigantopithecus,* so maybe it's been a rare animal all along. The evidence is in-conclusive. I think that it's either a very elaborate hoax or else a reality."

The mystery of Bigfoot acquired new substance in 1972, when biologist John Mioncznski introduced some astounding evidence. He had come upon some strange hairs snagged on a fence in the Wyoming mountains. Specialists believe that the strands are not from native wildlife and are probably primate hairs of a yet-undetermined species. Despite such find-ings, the majority of the scientific community remains extremely skeptical of Sas-quatch. "Sometimes scientists are the last people to admit to anything new in science; not very long ago we doubted the existence of mountain gorillas in Afri-ca," John said as we rode to Arrow Mountain on a pack trip into the Winds. John had participated in a major grizzly study in the area. He used radio telemetry to track the species, whose range in the lower forty-eight has been reduced to re-mote mountain regions of Wyoming and Montana.

"I noticed bears show far more individual behavior in their eating habits than other animals. For example, some grizzlies never try to fish and will jump right over a creek. Others head for a stream and become totally absorbed in catching trout," John told me. "It's a matter of training, based on the fact that a cub spends such a long time with its mother."

The study gave John his first clue that some bears seek out medicinal plants to cure ills. "I've observed that grizzlies gorge themselves on green gentian just after they have come out of hibernation. Now there's no food value in this plant, which pharmacologists cite as an extremely harsh purgative. Furthermore, early spring is the only time these animals have been seen feeding on green gentian. Clearly the bears are deliberately cleaning their systems after hibernation.

"My own goats select fringed sage when I've known they were suffering from worms, and that's the only time they do." John began using goats as pack animals when he was involved in research on Rocky Mountain bighorns. "The goats negotiate the high rocky ridges where horses can't go. And they don't spook the sheep. A castrated billy goat, known as a wether, carries 70 pounds. A nanny carries half that, but she'll give you a gallon of milk a day," said John. "Most goats are terrified of water, but my portage goat, Heidi, will walk right down the center of a canoe with her panniers on and then not move a muscle till we reach shore."

Our horses had been carrying us along a trail rising above the Wind River. When we reached the shade of a pine forest, John stopped often to talk about the plants along the way. I learned that common yarrow stops infection and hastens the healing of wounds. Arnica's crushed leaves applied in a poultice relieve pain. The leaves of mountain bluebell taste remarkably like green beans, and miner's lettuce makes excellent salad.

On the second day the forest became sparser and the trees smaller, until we reached the zone where stunted spruces and firs hugged the ground. They had been pruned by the savage climate into gnarled and twisted shapes. These elfin trees, called krummholz from the German terms meaning "crooked" and "wood," tell of years of wind and cold and driving snow.

Just beyond the krummholz is the sky-filled world above tree line. Such highlands are the summer range of the Rocky Mountain bighorns. Human encroachment on their rangeland and diseases from domestic livestock have driven the bighorns from many pockets in this highland wilderness.

Bighorns belong to the large group of incredibly agile hoofed animals that move through a no-man's-land of knife-thin ledges and steep cliffs with remarkable ease. In Europe, the chamois' gravity-defying leaps in the Alps and Pyrénées are legendary. A relative of the American mountain goat and actually a goat-antelope, the chamois can broad jump a chasm 20 feet wide and bound 13 feet straight up from a standing start. In addition to powerful muscles for launching the takeoff and a sturdy bone structure for withstanding the shock of landing, the chamois relies on specialized cloven hooves that provide a suctionlike grip. Sharp dewclaws on the ankles help prevent skidding.

The world's greatest variety of wild sheep and goats roams the Himalayas, Karakorams, and Hindu Kush—the stupendous heights that separate the Indian subcontinent from the Tibetan Plateau and Central Asia. These animals have such uncommon names as argalis, markhors, tahrs, and bharals. The research biologist George B. Schaller spent three years studying these mountain ungulates—known to few outsiders.

To look for markhors, shaggy wild goats with long, corkscrew horns, Schaller went to Pakistan to the Hindu Kush Range, a harsh, angular world. He saw several markhors—"faint brownish dots lost in the immensity of rock." Through a spotting scope he observed the animals "traversing ledges so precarious that footholds seemed more imaginary than real. . . . Poets may praise the deer and nightingale; I celebrate the wild goat," asserts Schaller in his book, *Stones of Silence.* "A markhor male standing *(Continued on page 74)*

*M*olten rock flows down Italy's Mount Etna, where legend says Zeus buried a fire-breathing monster. During this major eruption in 1983, a volcanologist (opposite) scoops out lava to analyze trapped gases. Another huge blast by Etna in 1985 leveled a resort hotel, killing one guest and injuring others.

PAGES 68-69: In the French Alps, summer climbing parties tread an icy ridge on Mont Blanc, the tallest peak in Western Europe. The Alps, geologists say, began rising 25 million years ago when northward-moving Africa compressed Europe's midriff into high, accordion-like ridges.

PAUL CHESLEY (68-69)

*I*n one of the least penetrable regions on earth, a porter negotiates a makeshift
bridge across a narrow chasm. Eighty feet below courses the Braldu, a glacier-fed
river in the Karakoram Range—a rugged spine in the Kashmir area of the
Himalayas. Kayaker Bob McDougal (opposite) threads the narrows of Braldu
Gorge in 1984 to become the first to navigate the river's 55 violent miles.

PAGES 72-73: Kim Schmitz in 1983 tries to be the first to scale the west ridge of
Mount Everest without using bottled oxygen. Hundred-mile-an-hour winds and
the thin air near the summit of the world's tallest mountain—$5^{1}/_{2}$ miles above sea
level—turned him and his climbing party back less than a mile short of their goal.

on a promontory, ruff shining in the sun, is a far different goat from the small, disheveled, smelly beast that domestication has produced."

The search for ibex and Marco Polo sheep took Schaller to the Karakorams. This range contains some 20 summits that tower above 25,000 feet, including K2—the world's highest peak after Mount Everest. He hiked past glaciers and "eroded slopes that looked as if a monstrous cat had clawed them in fury." He noted that remote villages seemed mere spots of green amid the sheer, naked rock. On a plateau at 15,600 feet he stood with one foot in Pakistan and one in China. "A ridge to the west belonged to Afghanistan, and beyond, across a valley, the snow-bare slopes were in Russia. The only sign of life was a fox track."

Schaller eventually found the ibex but not the sheep. Some have been killed by trophy hunters, but most have been cut down by local people for food. Like the sheep, the other highland ungulates are threatened and are making a last stand. Schaller reminds us that without these Pleistocene survivors, which can thrive in an ice-age realm of glaciers and rock, the mountains will turn into "stones of silence."

It's hard to imagine what life at the top of the world would be without these animals, especially the yak—a squat, shaggy bovine. The female gives milk for the butter that flavors tea and burns in the votive lamps of Buddhist shrines. The yak provides hair for clothing and dung for fuel in a land with little wood. When they die, they yield meat, and their tails are exported to India for use as ceremonial fly whisks. Add to all that a lifelong service of pulling plows and carrying cargo. More nimble and reliable than a horse, this beast can cross the highest passes belly-deep in snow, and in the Himalayas, on treacherous trails that skirt abysses, the yak plods stolidly along. Just don't take it below 10,000 feet, for then, as the Sherpas say, "Yak get very sick."

The monumental beauty of the Himalayas and the magic of just being around Mount Everest (at 29,028 feet the mightiest of mountains) have drawn another species—trekkers—to the dizzying heights. The trekkers come from North America, Europe, Australia, and Japan. About 25,000 a year journey to Nepal to walk along foot trails and follow ancient trade routes. Sherpas serve as guides and porters to carry food and equipment.

Nothing prepares you for the setting. Unearthly white mountains stand sentinel over rhododendron forests. Pink and scarlet blossoms frame views of glittering summits. Red-clad Buddhist monks sound horns that send low haunting wails for miles across an ocean of high ridges. The clarion air brings immense vistas into sharp focus, telescoping vast distances and dwarfing everything but the peaks, the glaciers, and the sky.

Many trekkers head for Sagarmatha (Mount Everest) National Park in the heart of the Sherpa homeland and set out on the trail to the foot of Everest. Breathing hard and strangely lightheaded, they head for Kala Pattar, a vantage point at 18,200 feet. It offers a sweeping view that includes the giants Lhotse and Nuptse, and the savage Khumbu Icefall.

No one can predict who is likely to suffer from serious altitude sickness. Neither age, nor sex, nor physical fitness is a factor. A headache, wooziness, and shortness of breath plague nearly everyone who flies into the 9,400-foot-high airstrip at Lukla, then trudges up another 3,000 feet or so to Namche Bazar, a Sherpa village and the entrance to the park. Those who too quickly go even higher risk more ailments—nausea, insomnia, and even pulmonary edema, sometimes fatal within 48 hours. But with a little time we lowlanders adapt, and the fatigue and symptoms diminish in a few days as our breathing becomes more normal.

"However, you'll never match the Sherpa's ability to work at high altitudes," Dr. Paul T. Baker, anthropologist at Pennsylvania State University, told me. "For a long time people assumed Sherpas had an unusually high red blood count and large lungs. Tests show that is not the case. We're not sure how, but in some way the Sherpa is able to extract more oxygen from the air. I—and many others— have spent 15 years trying to understand how it is done, but it remains a mystery."

Horses collapse and mice fall into a coma at sudden, extreme changes of altitude. But some birds don't seem bothered at all. Mountaineers on Everest have seen ravens fly up to 22,000 feet to pick at provisions left behind by past expeditions. Even the thin air of three-mile heights doesn't stop snow leopards, rarest and least known of the world's large cats. These agile hunters with large padded paws leap from boulder to boulder as they stalk mountain sheep and goats. Other creatures also find sustenance in that frozen world at the edge of the sky. Populations of tiny insects and daddy longlegs feed on pollen, seeds, dead insects, and plant fragments borne aloft by the winds. This community of life exists in what has been dubbed the aeolian zone, after Aeolus, the Greek god of the winds.

While windblown debris attracts insects to the edges of the earth, something else summons a special breed of human to climb the Himalayan peaks. At 24,000 feet and above, tying shoelaces is hard work, and each labored step takes several breaths. Winds cause small hemorrhages in the eyes, and the dry air evaporates body fluids. Crevasses often lie hidden beneath the snow, and avalanches may blast down the slopes and sweep a climber to his death at any time. What, then, is the lure of Mount Everest?

"Sheer joy," says Barry Bishop of the National Geographic Society's staff. In 1963 Barry was a member of the first American team to scale Everest. "You climb for the joy of extending yourself mentally and physically to the edge . . . for the joy that comes from the camaraderie and teamwork required to meet the challenge . . . for the joy of working with the cheerful and selfless Sherpas."

In 1978 Italian Reinhold Messner and Austrian Peter Habeler became the first climbers to attain the summit of Everest without using bottled oxygen. Reinhold finds a spiritual exhilaration in climbing. The important thing, he says, is how man copes with the critical situations, with solitude, with altitude. "My expeditions have thus enabled me to draw closer to myself, to see into myself more clearly," he notes.

For Herbert Swedlund, who has scaled steep rock walls in Yosemite and who has been a climbing guide in the Tetons for 25 years, part of the lure of mountaineering is being at the edge of uncertainty. "The willingness to accept risk for immaterial reasons is a very high human attribute," he told me. "After so many years of guiding I've learned to spot a good climber. Mental attitude counts as much as good balance and a high strength to weight ratio. In climbing, a person can't step around or avoid difficulties. He has to be able to exhort himself, to create energy in himself for the task ahead."

Mountain climbing as a sport was born some 200 years ago in the Chamonix Valley in the Alps. From there the wealthy launched their assaults on Mont Blanc, Western Europe's highest peak. They set out in large caravans, which in one case included such equipment as mattresses, bed linen, a white suit, five pairs of shoes, and a parasol. One countess reached the summit decked out in attire that included fur gloves and a boa. I did not covet the boa, but a good sturdy umbrella would have been welcome during my climb of Mount Kinabalu on Borneo. At 13,455 feet this peak is the highest mountain between the Himalayas and New Guinea. Here tropical rains average 116 inches a year. I wore a slicker against the predictable heavy downpour, but rivulets ran from my back down my legs, and my waterproofed hiking boots squished when I walked.

I had come to Kinabalu to see what I could of one of the richest and most remarkable assortments of plants on earth: the *Rafflesia keithii,* with blooms a foot and a half wide; begonias five feet tall; luminescent fungi that glow at night like faint neon lamps; a thousand species of orchids, many of them found no place else; and *Nepenthes rajah,* a giant among pitcher plants with a bowl that can hold a quart of liquid.

With Gabriel Sinit, a senior ranger in Borneo's park system, I saw some of these wonders, which provide many excuses to stop on the steep trail that ascends 7,000 feet in five miles. In the evening, after we reached the hut shelters at 11,500 feet, the rain stopped; the clouds parted and through a veil of pink mist loomed a granite escarpment with great rock pinnacles. Kinabalu's upper reaches resemble the fortress spires of some unearthly kingdom as mystical and haunting as a disturbing dream.

The next morning at three o'clock sharp we set out in the clear, cold darkness, using flashlights to find the trail. It became an internal struggle to force myself to go fast enough to reach the summit before the clouds relentlessly closed in. Finally there was nothing above but the midmorning sky. Below me lay Borneo's immense rain forest, a remnant of the vast tropical forest that once covered much of Southeast Asia.

The green canopy blanketed the rolling mountains of the Crocker Range, and beyond stretched the South China Sea. The view was utterly splendid, enhanced a little, I think, by my elation. Like any person who has ever scaled a peak, whether the Matterhorn of the Alps or Old Rag of Virginia's Blue Ridge Mountains, I was savoring my achievement. But the challenging journey to the top was more than an athletic enterprise; it was one more step in the discovery of the wonders of the mountain world.

Fortress in the clouds, Mount Roraima soars more than a mile high. Its sheer sandstone flanks wall off Amazonian rain forests on the border of Brazil, Venezuela, and Guyana. Geologists speculate that the mesa, or tepui—discovered in 1835—once formed part of an immense plateau in Africa before South America broke away 100 million years ago. Some scientists say that a species of frog similar to one in Africa inhabits Roraima's "lost world" environment.

ROY W. MCDIARMID

*M*ystery species abound on 10,000-foot-high Cerro de la Neblina—Mountain of Mist—a remote plateau in Venezuela. Scientists have yet to classify the tree frog (below, right). The purpose of its eye stripes presents a puzzle for scientists. Found only on Neblina, shrubs called Neblinaria celiae *resemble long-stemmed, wide-open artichokes (below, lower). Their bark, as thick as two inches, helps protect them from fire, which may flare from lightning strikes during periods of drought. Beauty in the bogs, Neblina's pitcher plant,* Heliamphora tatei *(below, left), fills its protein needs with insects that fall into tube-shaped leaves holding pools of digestive liquid. Flag-pole palms,* Geonoma appuniana *(opposite), mark the summit of a peak that rises abruptly from the 50-mile-long mesa. Botanists believe that many plants on the isolated plateaus in the Neblina area evolved there and grow nowhere else.*

ROBERT W. DICKERMAN/AMERICAN MUSEUM OF NATURAL HISTORY (ABOVE, LEFT); ROY W. MCDIARMID (ABOVE, RIGHT)

MERCEDES FOSTER (OPPOSITE); THOMAS GIVNISH, PH.D (ABOVE)

Glaciers and snowfields cloak the peak of 22,205-foot Nevado Huascarán, the tallest mountain in the Peruvian Andes. The Andes began to emerge about 25 million years ago when the Nazca plate, an eastward-moving slab of the Pacific seafloor, started diving under South America. Still squeezing underneath the continent today, the leading edge of the plate bends toward the earth's fiery heart and dissolves into molten rock, or magma. The buoyant liquid constantly pushes the overriding land skyward, breaking and shifting layers of solid rock. The Andes move upward at the rate of about one foot every 300 years.

Summer moon in Wyoming floats above icy ridges of Buck Mountain in Grand Teton National Park. Youngest and smallest branch of the Rockies, the Tetons started rising about ten million years ago when movement along a fault line thrust up rocks some $2^{1}/_{2}$ billion years old. Ice Age glaciers later chiseled the jagged spires and scooped valleys, such as the basin below. It cradles Lake Solitude within sight of 13,770-foot Grand Teton, the park's highest mountain.

FOLLOWING PAGES: Far beyond the giant outcroppings in the foreground, hazy Mount Whitney caps California's Sierra Nevada with the highest peak in the contiguous U.S. Through the ages, expanding magma and powerful quakes have added the jumbled granite blocks of the Sierra Nevada to the mountain world.

Emerald Kingdoms

The Forests

By Jane R. McCauley

*Highland waters of Parijaro
Falls plummet more than 800
feet to the floor of a limestone
amphitheater in the Peruvian
Andes. The light aircraft to the left
of the cataract seems little larger
than a jungle bird above the
tropical forest's canopy. Monarchs
of the plant kingdom, trees
number among their kind earth's
tallest and largest living things.*

*FOLLOWING PAGES: A dozing
sloth hangs from a branch in a
Casearia tree. With low body
temperatures, these slow-moving
creatures wear fur coats to keep
warm, even though they live in hot
Amazonian forests.*

*"H*ot lips," *the common name for the shrub* Cephaelis tomentosa *(below), displays bright, puckered bracts that attract pollinators in a Panamanian rain forest. Flashy bracts of the Amazonian herb* Heliconia *(opposite) lure hummingbirds.* Heliconia *flowers fit the shape and length of the beak of the particular kind of hummingbird that pollinates them. Lower: Lacking flowers, seeds, or fruit, the darker climbing fern* Salpichlaena volubilis *and the lighter tree fern* Cyathea trichiata *rely on the wind to scatter their spores.*

The Forests

Walking early on this morning in the midst of the forest, I find the land quiet, almost empty. This place seems different, unlike any woods I have seen. Trees stand taller than a hundred feet. Vines dangle from limbs. Near me I notice a palm, its trunk covered with prickly spines. Alone in the quiet, I could feel wary. But there is a reticent beauty to the forest, a curiosity that envelops me instead. Unaccustomed to the tropical heat and humidity, I move slowly. Little breeze stirs under the trees. Skies have cleared after last night's downpour. The soil beneath me seems to have absorbed most of the water. It is early June, and the rainy season has begun. An average of 100 inches can fall every year, most of it between May and December.

Suddenly a rustling on the ground breaks the silence. Out from the leaves jumps a toad. A flash of emerald reveals a parrot flitting among the treetops. Howler monkeys let out their deafening roars.

Here, in a tropical forest on Barro Colorado Island in Central America, I began my journey to learn more about the forest realm—for me a fascinating and unpredictable world. From low-lying woodlands and mountain cloud forests in Panama and Costa Rica to the rich cove hardwood forests of the southern Appalachians, I encountered amazing plants and animals and the clever survival mechanisms they have developed over eons of trial and error. I confronted some intriguing theories about the origin of certain forests. I was amazed by the incredible array of life sustained by rain forests of the tropics. I found it interesting that certain related species of trees grow wild in the United States and in China but not in many of the lands between. And in Japan I became absorbed in the culture of bamboo, a plant that grows so fast that you can actually watch it shoot up.

Forests cover approximately one-fifth of the world's land surface in three major belts: boreal, temperate, and tropical. Of all the forests I visited, the most curious and challenging to me were the steamy rain forests. They flourish close to the Equator, where temperatures can average close to 80°F, and precipitation can be as much as 400 inches. The lush domain of tropical forests—accounting for roughly a third of the world's forests—cuts a swath nearly as large as Europe through Latin America, Asia, Africa, and across the tip of Australia.

"It is by its sounds that the life of the forest makes itself known to the outsider," reads a passage from Catherine Caufield's *In the Rainforest*. "The disquieting roars of howler monkeys fade out, to be replaced in turns by cries of unseen birds, the crash of a falling branch, the rustling of leaves as a small creature scuppers away, and other, unidentifiable, noises. The newcomer sees little movement or color. It helps to be quiet, to have an eye for detail, and to know where to look; it is even better to have a companion who knows the forest well."

My experiences on Barro Colorado seemed to echo Caufield's observations. This 3,700-acre forest in Panama covers an island that serves as an outdoor laboratory run by the Smithsonian Tropical Research Institute for scientists from all over the world. Biologist Alan Smith guided me through the forest, perhaps the most studied in the world. I welcomed his expertise. He is aware of the need to increase the public's knowledge of tropical forests. Through him, I discovered that these forests are exciting, not the unfriendly and impassable jungles often portrayed by movies and novels.

Before beginning our walk, I sprinkled my ankles with sulfur powder, then taped my pant legs, a procedure sometimes used to ward off chiggers and possibly the ticks that abound here. As we entered the forest, spider monkeys cavorted on branches near us. Alan told me that the forest can look different at various times of the year. He explained that seasons do occur in the tropics. I understood from him that these are not the spring, summer, winter, and fall of the temperate zone. In the tropics rainfall influences the seasons. "Barro Colorado is a forest with marked dry months, typically January through April. About 15 percent of the tree species drop their leaves during this time, opening the forest," he told me. He also mentioned that there are plants and animals that appear to go dormant. Where rains fall more evenly throughout the year closer to the Equator, the seasons are not as pronounced.

Moving through deep shade and past tall trees, I was struck by the monochromatic look of the forest. At first I thought much of the foliage looked alike. Traveling in the Amazon in the mid-1800s, naturalist Alfred Russel Wallace made note of such muted colors: "There is a grandeur and solemnity in the tropical forests, but little of beauty or brilliancy of colour. The huge buttress trees, the fissured trunks, the extraordinary air roots, the twisted and wrinkled climbers, and the elegant palms, are what strike the attention and fill the mind with admiration and surprise and awe."

Like most newcomers, I thought vegetation at ground level would be impenetrable. I found out quickly that this is a myth. Dense, low vegetation can occur along sunny riverbanks or in patches where a tree has toppled and left a hole in the canopy for more light to pour through. Actually some of the thickest growth is found overhead. In many parts of tropical forests, vines lace through the trees, and other plants in assorted hues of green often layer trunks and branches. Some rainwater, soaked in by these plants, may never reach the ground. "So diverse is the world of the forest canopy," says tropical expert Norman Myers, "that it can be considered the last great frontier of biology. No other habitat sustains a greater abundance and concentration of species, both plants and animals."

One of the most noticeable things for me in Barro Colorado's forest was the variety of ants. Early in our hike, Alan pointed out leaf-cutter ants marching in precision only a few feet from us. Each one carried a green morsel. Alan showed me an area several feet across that they had stripped of foliage. "They take the leaves back to underground chambers, chew them, then deposit them to fertilize the fungus gardens that give them their food," he said. He explained that leaf-cutter ants shun certain leaves because they are toxic to them. In the past few years scientists have recognized that the ants also stay away from plants toxic to the fungi. "From some of the plants they avoid," Alan indicated, "scientists may derive a compound to cure skin fungi in humans, such as athlete's foot."

Many tropical forest plants have medicinal properties. Norman Myers writes: "When we pick up a medication . . . at our neighborhood drugstore, there is roughly one chance in four that the product owes its existence, either directly or indirectly, to raw materials from tropical forests." Curare, for example, comes from several poisonous plants. This extract contains alkaloids that can make muscles relax. In the field of medicine they can help reduce the amount of anesthesia a patient needs during surgery. Some South American Indians still use curare in hunting to paralyze game. They coat tips of their blowgun darts with curare. The slightest prick of the skin by one of the darts can prove fatal.

I paused a moment and took a close look at the climbing, woody vines—or lianas—looping through the treetops. The vines rise from the forest floor, wrapping around trunks. Too tight a grip can strangle a tree. Rattans are some of the longest of these vines, growing to lengths of 500 feet or more as they weave through the treetops, competing with trees for space and light.

I wondered how plants here could survive in the lower light of the forest. "The trick is either to get as much light as possible or to adapt to the deep shade," Alan said. Certain plants have leaves with movable joints in their stems to capture shifting rays. Pointed tips in other plants shed water rapidly, most likely preventing a buildup of light-absorbing mosses and fungi. "Some sunlight reaching plants is absorbed by the leaves," he explained, "while some light often passes through them. A few species have developed leaves with red undersides that help trap the light inside, preventing its loss."

I turned my attention back to the trees. It struck me how much easier those at my home in Virginia would be to climb. Here in the tropics, the limbs of trees often begin several feet off the ground. No matter where I looked, it seemed I was looking at a different kind of tree. The amount and diversity of life in tropical forests can be overwhelming. Though well concealed, around 1,350 species of plants, 369 types of birds, and 65 kinds of mammals fill Barro Colorado alone. And the numbers are far greater in wetter equatorial regions. While an acre in the Appalachians may hold some 20 kinds of trees, a given patch near the Equator can have 5 times as many. "Why are these forests so profuse?" I asked. Alan answered, "We do know it's not because they are stable. We used to think so. Now we know they are among the most dynamic communities on earth."

I thought about the many months it would take to know Barro Colorado, so teeming with life. The forest had quickly taught me that it reveals itself only at its own pace. How difficult the challenge, then, of probing even larger forests, such as the rain forest of the mighty Amazon basin.

Stretching out across South America, the Amazonian rain forest—the most extensive of its kind in the world—encompasses some two million square miles, an area about ten times the size of Texas. Meandering through the Amazon basin's verdant maze, the river that gave the region its name is—by volume—the largest river in the world. Nicholas Guppy, a tropical forest ecologist, has known this area since the early 1930s. "Today," he said, "outsiders have colonized much of this wild region. You can take a boat 2,800 miles up the Amazon River, and you won't see virgin forest. But when you do find it, it's incredibly beautiful, dark, absolutely breathtaking in structure."

The best way to see the vast Amazon is to fly from eastern Colombia across Brazil. From the air the mushroom-shaped trees merge into an endless vista of green. Waterfalls cascade hundreds of feet, and giant lily pads float in stagnant pools. Outspread arms of inundated trees poke from rivers illumined by the fiery sun. How parts of the Amazonian forest withstand torrential flooding during the rainy months each year remains one of the area's many enigmas.

The variety and number of life-forms here challenge one's imagination: a million different kinds of plants and animals, among them orchids and mammoth rose and violet trees; at least one out of five of every known bird species, including rainbow-hued macaws and toucans; perhaps 2,000 kinds of fish, counting the fierce piranhas; countless insects and spiders, one a hairy tarantula big enough to prey on birds; the capybara, largest known rodent; and the giant anaconda, one of the earth's largest snakes.

"The Amazon's like a museum in perfect order," Guppy says. Ingenious alliances between plants and animals help ensure the steady rhythm of life. Tiny brown-and-yellow orchids gleaming in the sunlight suggest swarms of bees. Male bees in the area mistake them for rivals and attack, spreading the flowers' pollen. Fruit-eating bats are especially drawn to one particular type of fig tree. Its flowers and fruit grow on its trunk rather than on its branches—a phenomenon known as cauliflory. The low position of the figs, scientists think, helps bats avoid snagging their fragile wings on the limbs while they are feeding on the figs at night. In return for accessible food, the bats distribute the seeds, thus helping ensure the survival of the fig species.

Paradoxically, the soils that support the Amazon's trees and plants are some of the most infertile in the world. "These are fossil soils," Guppy said. "The soils are so poor simply because they are so old. Over millions of years, rainwater has been leaching away their minerals."

"How, then, can this region be so lush?" I asked. "Because of diminutive fungi called mycorrhizae," Guppy replied. They are the lifeline of the Amazon basin. Roots alone aren't able to extract enough food supplies from the deficient soil. That's where the threadlike fungi help out. They develop on tree roots, and they help roots collect nutrients, which are then stored in the trees and plants; in temperate areas, nutrients are stored in the soil. In the Amazon's remarkably self-sufficient system, nutrients recycle from dying plants to living ones. When a leaf tumbles to the ground, the minute fungi quickly snare it. The mycorrhizae are highly efficient at gathering and channeling phosphorous and other minerals to the tree. "Once these tiny fungi are lost or damaged, the trees won't grow back," Guppy added.

Whereas the Amazonian rain forest owes its existence largely to the tiny mycorrhizae, a cloud forest in Costa Rica probably owes its preservation more to a rare toad than anything else. A narrow path led me into the Monteverde Cloud Forest Reserve, an 8,000-acre sanctuary in the Tilarán Mountains. The tall trees, rising through swirling mist, resembled spires in a fairy-tale kingdom.

Cloud forests are named for the low, moisture-laden clouds that wrap them in mist nearly year-round. These forests often cloak mountain heights in Central America and on Caribbean islands, where soothing trade winds nurture luxuriant vegetation. The forests are remote and usually out of reach; one of Monteverde's attractions is its relative accessibility. Each year, hundreds of nature enthusiasts hike along the path that I was following.

I had never seen such rampant growth: ferns as *(Continued on page 101)*

*G*ossamer mist wafts through
an enchanted woodland, the
Monteverde Cloud Forest Reserve
in Costa Rica. Here, on wind-
buffeted peaks that straddle the
continental divide, more than
2,000 plant species compete for
survival. Epiphytes—clinging
plants that include ferns, mosses,
and liverworts—climb the stunted
trees and absorb nutrients from
the air. Heavy condensation from
moisture-laden trade winds helps
make Monteverde the most
luxuriant cloud forest in the
American tropics. Creatures of
fear and fantasy dwell within the
reserve's boundaries. A bird
prized by the Aztecs, the
jewel-colored quetzal flashes
among the branches. Sure-footed
ocelots prowl the underbrush.
Tiny golden toads—found
nowhere else—emerge to mate in
puddles that dot the forest floor.
The discovery of these rare
amphibians in 1964 helped win
protection that keeps Monteverde
a wilderness area.

CAROL HUGHES

Camouflaged or conspicuous, tiny New World tree dwellers discourage predators by blending with the environment or by advertising themselves as unpalatable morsels. Allergen-filled spines protect a stinging caterpillar (below), known as a "little cypress" because its prickly hairs look like evergreen trees. Clustered in defense, the saddle caterpillars (opposite) will develop similar syringe-like hairs, as well as a bull's-eye pattern that warns of their bitter flavor. Leaf-green bodies help a cone-headed katydid (lower, left) and a red-eyed tree frog (lower, right) escape detection.

CAROL HUGHES (ALL)

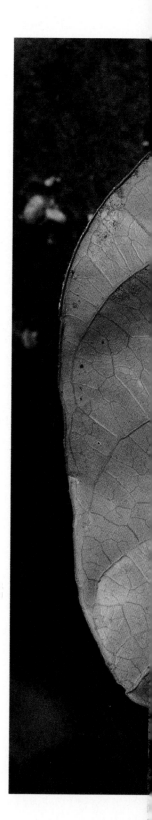

tall as trees; carpets of thick mosses; royal blue butterflies nearly half a foot long; huge lichens, shaped and ridged like clamshells; and countless epiphytes, or air plants. Nearly every inch of soil was covered. Spiderweb nets hung between clusters of leaves in the all-encompassing shade. Monteverde fulfilled my image of true jungle.

Epiphytes such as orchids and bromeliads are so voluminous here that a single tree can wear a hundred different plants, draped like layers of clothing. Epiphytes anchor themselves to their hosts, sucking in nourishment from rainwater and debris. One of them—the genus *Blakea*—has species that differ from all other plants in the New World. Wilford "Wolf" Guindon, a member of Monteverde's Quaker community, told me about one of the species as we headed along a path to see the forest's special toads.

Blakea chlorantha, a green-flowered plant, opens and secretes nectar at night, Wolf said. For a long time no one could figure out what pollinated it. New York botanist Cecile Lumer spent many nights in the reserve before she solved the riddle. She saw a mouse scurrying up a trunk to drink the nectar. Her finding in 1980 was the first recorded sighting of a plant being pollinated by a rodent in the Western Hemisphere.

Our journey took us past spiky bamboos, clattering palms, and quiet waterfalls. In a fairyland like this, the most natural sounds are spooky: the drip, drip of water; a distant volcano's rumblings; a howler monkey's roars; the clang of the brown-and-white bellbirds, loud and clear as church bells on a Sunday morning. Perched quietly in the recesses of the forest is the resplendent quetzal. The male of this endangered species is distinguished by its long train of glittering emerald tail feathers. The ancient Aztecs pictured their god Quetzalcoatl adorned with the bird's vivid plumes. Quetzals nest in large numbers in Monteverde, drawn here mainly by the fruit of laurel trees.

We crossed the windswept ridges of the continental divide, where the tall trees gave way to the stunted ones of an elfin forest. We were getting close to the forest's celebrated amphibians. Afternoon clouds closed in. Wolf beckoned me to a nearby tree. First I heard their soft trilling. Then I saw one. Then another. Then six or seven more leaping out of a murky puddle. The orange color of Monteverde's rare golden toads gleamed like patent leather.

"Females wear more muted colors, usually olive or black with scarlet dots," Wolf said. Biologists think that the bright coloration of the males attracts the females. But for some inexplicable reason, the males cannot recognize them. At mating season they simply clasp any toad that comes along until they finally latch onto an appropriate female.

Native of New Guinea's rain forest, a male lesser bird of paradise exhibits the elaborate plumes it flaunts to attract a female. During a complex courtship ritual, as many as ten males dance at the same time in a single tree. A suitor raises its flank feathers and, rocking back and forth with wings spread, lowers its head and lets the feathers cascade over its back like colorful streamers.

THE STOCK HOUSE/LARRY BURROWS

Wolf had helped research biologist George Powell win protection for Monteverde and its distinctive assortment of wildlife in 1972. Powell rallied organizations and individuals worldwide to the defense of the golden toads—which live nowhere else in the world.

Tropical forests have kept many natural treasures secret for centuries. The golden toads weren't discovered until 1964. In La Selva, a rain forest northeast of Monteverde, I observed an unusual species of flying mammals. Early one morning, I accompanied bat specialist Anne Brooke into the jungle. In the midst of shoulder-high foliage, she pointed to a broad umbrella-shaped leaf. On its underside, clinging to a thick vein, were four white bats the size of marshmallows with eyes black as raisins. In six weeks she had spotted 40 of the animals, whose range covers a narrow zone in Central America from Honduras south to Panama.

In Australia another group of interesting mammals—the marsupials—can be found in coastal forests. Using their elongated forelimbs, tree kangaroos leap from branch to branch. With folded skin that fans out to form a "flying carpet," the greater glider—Australia's largest gliding marsupial—sails between gum trees; its "flights" can cover the length of a football field. Teddy bear look-alikes, the koalas nap in the crook of eucalyptus trees. Australia has more than 500 species of eucalyptus trees, but the finicky koalas eat the leaves of fewer than 12 of them. Why? No one knows for sure.

Thriving in hot, dry environments, some eucalyptus trees deviate greatly from most broadleaf trees. They have leaves that hang vertically, reducing water loss by exposing less surface area to the sun. Ordinarily, this loss of sunlight would reduce photosynthesis. But the waxy leaves offset this problem because cells containing chlorophyll cover both sides, capturing sunlight at various angles. Some eucalyptus trees are noted, too, for their lofty heights. But when it comes to standing tall, no living tree can match North America's coast redwood.

For some 500 miles along the Pacific coast of California and Oregon, the majestic redwoods can grow to heights of more than 300 feet. In the angular Sierra Nevada to the east, giant sequoias—some more than 100 feet in girth—stand as the earth's most massive trees. Although some of these evergreens date back several thousand years, the Great Basin bristlecone pines of Nevada, Utah, and California outrank them as the world's most ancient trees. Older than Egypt's Great Pyramid and gnarled by ages of exposure in an arid environment, some of the bristlecones have endured more than 40 centuries.

To the north at the Canadian border begins the boreal zone, an enormous belt containing the world's largest expanse of conifers. In North America, the boreal region reaches a width of 600 miles and sweeps about 4,000 miles across Canada and Alaska before tundra appears along the southern fringe of the Arctic. The boreal boundaries also take in parts of Finland, Sweden, and Norway, and the desolate reaches of Siberia.

A traveler in Canada's northern woods will find green-and-white bands of pines, larches, spruces, and firs rimming glacial lakes. Quaking aspens, poplars, and silvery birches nudge in between. The air smells of pine and damp mosses. So thick are the mosses in places that little rain touches the ground. Sometimes the land looks primeval, shrouded frequently by milky mists.

"The north woods of North America and Europe have about the same flavor in terms of weather," Dr. George La Roi, a forest ecologist at the University of

Alberta at Edmonton, told me. "In contrast, those of central and eastern Siberia have more intensely cold, dry, and cloudless weather."

Life can be harsh in Siberia. Here the boreal forest engulfs an area larger than the United States. Winter temperatures can plunge to a numbing minus 58°F and can soar above 90°F in summer. Only Siberian larches can tolerate such extremes and still dominate the landscape, Dr. La Roi said. Though needle-bearing, larches—unlike most conifers—shed their foliage at the onset of winter. "Thus," he said, "the trees reduce the damage they'd suffer from the biting winds and deep snows when the trees are frozen solid and are very brittle."

Around the globe, winter lingers in the far north and the growing season usually lasts no more than four months. Packed with glacial debris, the soil is scant. Rain, leaching acid from evergreen needles, seeps underground, dissolves minerals, and carries them away—leaving the thin soil even poorer.

"How do these trees survive?" I asked specialist S. C. Zoltai of the Canadian Forestry Service.

"Mycorrhizae help, just as they do in any forest," he answered. "These tiny fungi grow on the root tips, increasing the absorbing power of the roots so that the conifers take in adequate nutrients throughout the year. But what makes the boreal forest special is its adaptation to wildfires. Fires sweep across large tracts of the forest. In the period of a century few areas escape fire. Although the flames kill the trees, new forests spring up from surviving roots or from seeds that have been locked inside resinous cones for years. The heat of the fire opens the cones, releasing a rain of seeds and a rejuvenated forest."

Compared with the tropics, the northern woods harbor fewer species of plants and animals. Many of the creatures that do live here exhibit ingenious ways of handling the heavy snow. Under extreme conditions, a moose may use its forelegs as snowshoes, moving forward in a kneeling position for a short distance through a snowdrift or across ice. The willow ptarmigans grow feathers on their legs and feet. In addition to creating warmth by reducing loss of body heat, the feathers serve as snowshoes that help keep the birds on top of the fluffiest snow.

I was astonished when I found out how far south the boreal forest may have once extended in North America. "Just 10,000 years ago, at the end of the last ice age, glaciers had forced the boreal forest to migrate to the southern Mississippi River Valley and to the coastal plain of the Carolinas," Dr. La Roi said. "Before the Ice Age," he continued, "the northern forests of this continent were almost as rich in species as the present primeval forests of southeast China and the Great Smoky Mountains. In these areas the boreal conifers and hardwoods still dominate the higher elevations."

From mountaintop stands of evergreens to lower forests of broadleaf oaks, maples, and hickories, the Great Smoky Mountains embrace some of the lushest vegetation in the United States. More types of trees than are found in all of northern Europe, more than 100 species, and 2,000 different kinds of fungi grow in valleys and on gentle hillsides. Gray mists, which give the range its name, supply moisture to this abundance of flora. More significantly, hundreds of thousands of years ago the region escaped the life-threatening crush of the ice sheets. The fact that these forests have survived largely undisturbed (Continued on page 108)

G*raceful* moso *bamboo flourishes outside Kyoto, Japan. Scientists ponder the question: Is the plant a tree—or an oversize grass? It lacks a tree's growth rings, yet its hollow stalk, or culm, grows too tall and too thick for a grass. Snow can bend the plant; but, though extremely pliable, bamboo has super strength. One particular thicket survived the atomic blast that leveled the Japanese city of Hiroshima during World War II. Below, right: A hairy sheath protects a young shoot of moso bamboo. Savored as a tender delicacy during its first few days, the fast-growing shoot will reach its full height of 60 feet in less than three months. Occasionally culms develop abnormally, producing a variation known as tortoiseshell bamboo (below, left). Collectors value this rare aberration, so named because of the twisted culm's resemblance to a line of tortoises cowering head to tail.*

TAKAMA/PACIFIC PRESS SERVICE (ALL)

*T*aiwanese children on a
school outing explore a
forest of moso bamboo.
Many bamboo species
develop a subterranean
network of rhizomes, which
help hold the soil against
erosion. Abundant and
economical, bamboo
supplies Asians with
building materials for
thousands of products,
including roof tiles, siding,
irrigation pipes, fences,
baskets, furniture, and
flutes. For reasons that defy
explanation, some varieties
of bamboo flower only
every 60 or 120 years,
blooming and withering
simultaneously throughout
the world.

for centuries explains the bounty of trees in the Smokies. Or does it? Some botanists are beginning to wonder.

"In the 1970s we began to uncover evidence that some of these forests are actually far younger than we realized," Hazel Delcourt told me. At the University of Tennessee, I had met Hazel, a botanist, and her husband, Paul, a geologist. They have been reviewing the botanical chronology of the Smokies, mountains that run along western North Carolina and eastern Tennessee and form part of the southern Appalachians. "Thousands of years ago," Hazel said, "boreal trees and tundra made up almost all this area." Much earlier than that, Paul said, subtropical woods ruled. Hazel added, "Today, there's a tropical fern called *Vittaria* still flourishing under waterfalls." How did it cope with the drop in temperature during the Ice Age, I asked. "It may not have been here then," Hazel said. "The fern probably appeared here no more than 10,000 years ago. We aren't sure how it found its way here because we don't have direct fossil evidence to indicate when it established itself in the Smokies."

Another question confronting scientists concerns the richness of hardwoods in the Smokies. More than ten species fill cove hardwood sites at the lower elevations. Age has contributed to this wealth of species, ecologists have long believed; the hardwoods have had hundreds of thousands of years of uninterrupted time to diversify, they say. "But now Hazel and I think the forests were disrupted, pushed out by the cooler weather of the Ice Age," Paul said. By analyzing pollen from lake sediments, the Delcourts have concluded that oaks, maples, tulip trees, and other hardwoods retreated southwest to fertile upland areas along the Mississippi River. Hazel explained what may have lured them there. "Warm air from the Gulf merged with cool melting glacial water, creating a misty environment reminiscent of the present one of the Smokies."

"When it began to warm up about 10,000 years ago," Paul said, "the hardwoods began spreading back to the Smokies. This means that present-day forests are younger than previously suspected and that diversity is no clue to their age." Not all botanists, I was told, are ready to accept this new theory.

While wandering along mountain trails, many visitors in the Smokies come upon treeless spots, or heath balds. The abrupt transition from woods is startling, like stepping onto flat tundra. But balds are not tundra, since they fall below tree line. Why they are here is only a guess.

"Some naturalists believe the balds first appeared when settlers cleared the land for pastures," Hazel said. "Others suggest that wind, ice, or insects are the culprits. Some say the Cherokees made them by burning part of the forest. Even if they did, what is it that inhibits trees from reclaiming the balds?"

In the Smokies, I found a family of trees whose relatives live half a world away. Each spring, magnolias burst open their creamy white blossoms in the eastern United States and in China, habitats where these trees still grow wild. Millions of years ago, ice sheets and the breaking apart of continents killed off some members of this family. Consequently, surviving magnolia trees are glaringly absent today in many of the lands lying between the United States and China.

Because of a varied climate within expansive boundaries that encompass territory from the northern latitudes to just below the Tropic of Cancer, China offers a wide range of forest habitats. The evergreens in the brutal north, where the sun rarely shines, yield to deciduous trees, such as poplars, birches, and oaks. These give way to more evergreens in the monsoon forests of the south.

Farther west in the rugged interior province of Sichuan, forests of bamboo weather fierce winds and drifting snow. Engaging pandas live in this remote kingdom, content to spend their days munching shoots of bamboo.

Though most closely associated with the Asian countries, bamboo is widespread, occurring naturally on every continent except Europe and Antarctica. Some accounts credit merchants with distributing it along the ancient spice routes, winding through India, where today some of the largest reserves remain. I had first encountered this ubiquitous plant in the tropics, and in Japan I learned more about it.

On the islands that make up Japan, bamboo groves march down cliffs and rise on foothills far above workers tending rice fields in broad-brimmed hats. *Sasa,* a type of bamboo that sometimes grows only a few inches tall, crisscrosses the island of Hokkaido. The warmer islands farther south shelter varieties that measure 7 inches in diameter and stand more than 60 feet tall.

At Utsunomiya University north of Tokyo I called on Dr. Hiroshi Usui. He is one of Japan's most respected authorities on bamboo, and when he speaks of it, it seems almost sacred. "Bamboo is the main one of three plants that symbolize strength: the plum lives patiently through winter snows; pines survive in poor soils, and bamboo stays green year-round." Eventually, I would learn why Orientals also see in bamboo the virtues of humility and unselfishness.

Bamboo may well be the curiosity of the plant world. At least that is what one of the world's leading scholars on the subject, Dr. Koichiro Ueda, thinks. "Dr. Bamboo," as he is respectfully known, lives in Japan's ancient capital of Kyoto. On a narrow side street I located his home, where he welcomed me with traditional hospitality—green tea and a pastry wrapped in a sasa leaf. Dr. Ueda, spry for his 86 years, beams when he discusses his favorite plant.

"The more specialists discover about bamboo, the less they can agree how to classify it," he said. "Some label it a grass and put it in the grain family with wheat and rice. But others believe bamboo is a type of tree. After 40 years of research, I don't think it's either. I put bamboo in a category by itself."

"Bamboo is the world's fastest growing plant," he added. "I measured the record, a *ma-dake*—Japan's most common species—outside Kyoto. It shot up a yard in 24 hours. Most bamboos mature in two to three months by inching up both day and night."

Not like other plants, bamboos have hollow stems, called culms. "Never say stem," Dr. Ueda said. Stiff, yet uncommonly flexible when bent, the culm accounts for the plant's being a symbol of humility.

All varieties of bamboo propagate from underground rootstocks, or rhizomes. In old China, legend says, people gathered each spring at groves to await the audible popping sounds made by the young shoots. Out they come, at precisely the same width as the mature culms. Bamboo takes no more sustenance for itself after it has reached maturity; instead, the plant passes all its nourishment on to the young shoots it has spawned. This act of sacrificing its food led to the bamboo's legendary virtue of unselfishness.

Of the thousand or so varieties of bamboo in the world, Japan counts more than 600. I saw a sampling of them on the outskirts of Kyoto one sunny afternoon. Some species were square, some tall, some bent, and others squatty. Some carried frilly limeade-colored leaves that whispered in the breeze. Others bore tough leaves that thrashed about, moaning. One type especially caught my attention—the tortoiseshell. A thin crease zigzagged up its twisted culm, forming triangular bulges. What causes the mutation that distorts the culm and why it sometimes reverts back to the standard shape of the other culms are questions that scientists are trying to answer.

Though it may hide the reasons for many of its peculiarities, bamboo makes no secret of its versatility. I saw gutters, fences, ladders, baskets, fishing poles, table decorations, and, of course, chopsticks made of it. In China sturdy cables of bamboo suspend bridges. Bamboo scaffoldings in Hong Kong have withstood typhoons that have crumpled steel frameworks. Ironically, strong bamboo has a skin that scratches easily. It shows marks so clearly, in fact, that the Chinese used bamboo as paper before the second century B.C.

Nearly a century ago Thomas Edison, after failing with 6,000 other materials, found that charred ma-dake bamboo would glow when charged with electricity. Using ma-dake as a filament, Edison made his first successful light bulb.

Of all the oddities displayed by bamboo, none seems so bewildering as its flowering cycle. "Even the greatest experts cannot understand how the same species manage to bloom around the world in near unison," Dr. Ueda told me. Nor can experts explain why the intervals between flowerings are so long—30, 60, 120 years. The Chinese say the elderly are those "who have lived long enough to see the bamboo flower twice."

Its curious blooming cycle, its speedy growth, and its versatility are what make bamboo "absolutely the most mysterious plant on earth," claims Dr. Ueda. I'm sure others would agree. As for me, I've got to think about it for a while. After all, I had by now come across many bewitching forest wonders. I had seen rain forests overflowing with life, diminutive fungi making survival possible for huge trees at opposite ends of the globe, pines as old as the pyramids, bizarre wildlife, and plants that shifted from zone to zone to escape the deadly cold of the Ice Age.

In a nameless woods near my home in Virginia I sat down to mull all this over. I come here often to reflect on life's puzzles. Things never stay the same in these woods. Sometimes the trees or flowers look different. New animals seem to appear out of nowhere, and old ones vanish without a clue. But isn't that what my travels had taught me about forests? No matter how big or small, they're always changing, always a little mysterious, and forever fascinating.

In Washington's Olympic National Park a snowmelt stream spills past mosses and sword ferns. Thriving from Alaska to California, the Pacific coastal forest is one of the world's lushest woodlands. Marked by cool and moderate temperatures and the scarcity of woody, tree-climbing vines, North America's temperate rain forests contrast with the hot, humid, vine-choked ones of the tropics.

WILLIAM BOEHM

*C*arpet of conifers rolls toward Good Hope Landing, once a cannery and now a popular staging area for salmon fishing on the Sandell River in British Columbia. Fanned by moist Pacific breezes, the yellow and western red cedars cloaking these hills have escaped the commercial lumberman's saw for 250 years.

PAGES 114-115: *Reindeer thunder past larches in the Siberian taiga. This subarctic region of open woodland covers about one-third of the Soviet Union.*

*W*orld's oldest and largest living trees grow only in America's western forests. Rocky terrain suits Great Basin bristlecone pines in Nevada (below). These trees can live more than 4,000 years on dry, cold, windy sites. In the Sierra Nevada, giant sequoias (opposite) belong to a species noted for record girths exceeding a hundred feet. Each giant sequoia sprouts from a seed as tiny and thin as a flake of oatmeal.

PAGES 118-119: Sequoias of the sea, fronds of giant kelp reach out some 200 feet. In North America these submarine forests range from Canada to Mexico.

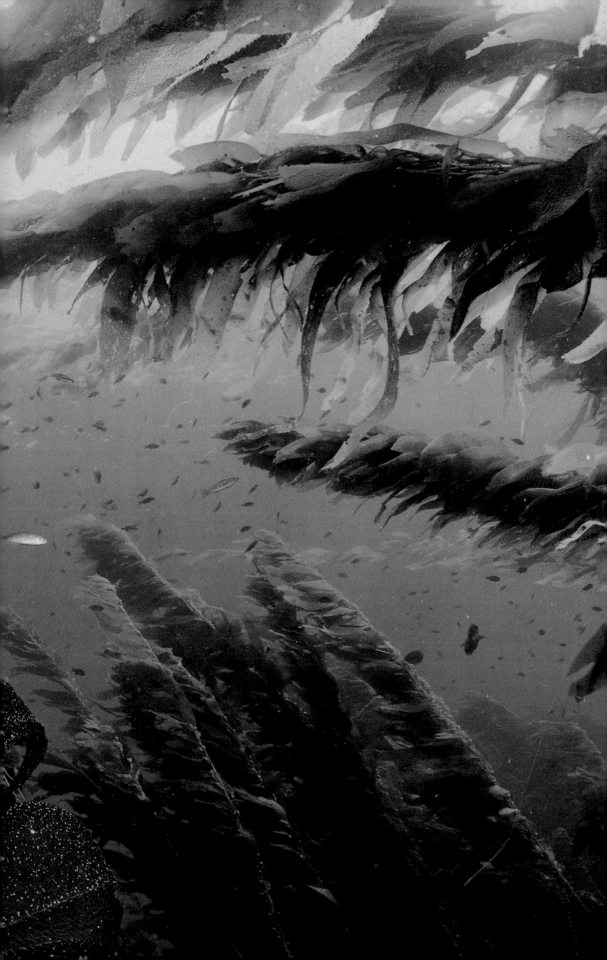

Plains Country

The Grasslands

By Christine Eckstrom Lee

Tawny coats the color of tallgrass camouflage lions in Namibia's Etosha National Park, one of many African sanctuaries that preserve wild grasses. Worldwide, grasslands fill the ecosystem between forests and deserts. On grassy plains man advanced from hunter-gatherer to pastoralist and farmer. Today grasslands support a chain of life ranging from microbes to mammals.

FOLLOWING PAGES: Wildebeest cross Kenya's Mara River. Here in the Serengeti 2.5 million grass eaters, such as the wildebeest, and 28 species of carnivores, including lions, coexist in the world's greatest assemblage of large wild animals.

*C*attle egrets shadow elephants in Kenya's Amboseli National Park; insects stirred up provide a feast for the birds. A single African elephant eats as much as 300 pounds of vegetation a day, and by uprooting trees as it feeds, it contributes to the spread of grasslands. Largest of the land animals, an adult elephant weighs up to seven tons and stands ten feet high at the shoulder. Born after a pregnancy lasting about 22 months, a young elephant remains with its mother until the early teens. Females share in caring for the young, occasionally nursing calves not their own, and are always ready to give a calf a reassuring pat with the trunk (opposite).

One reason grasses can triumph despite severe climatic extremes is that most of their living tissue is underground, protected from the world above. Some grasses have roots that reach depths of more than 20 feet. The root systems are often extensive. A single rye plant, for example, may spread nearly 400 miles of roots. Underground, countless species of insects play a critical role in keeping the soil rich for grasses. Africa's industrious dung beetles roll and bury balls of animal droppings that fertilize the soil; throughout the temperate zones, earthworms and rodents aerate the soil with their tunnels.

One acre of natural grassland is capable of supporting a greater mass of animals than an acre of any other kind of land. It does not seem surprising, then, that from grasslands came wild animals that became our domestic livestock, such as cattle and sheep. Grasslands also gave us our major grains: wheat, rice, oats, and maize. With the discovery of ways to domesticate animals and raise crops—more than 10,000 years ago—man began to change from a wandering hunter-gatherer to a pastoralist and farmer. This agricultural revolution encouraged people to settle in one place; over time, communities, cities, and civilizations formed—all nourished by the cereal crops and domestic animals that had their beginnings in wild, natural grasslands.

Today most of the food we eat is derived from grass. With vast acres of grasslands tamed for agricultural use, few of the world's grasslands still preserve their original varieties of plant and animal life. Poor farming practices and overgrazing caused a number of grassland species to vanish before we recognized the loss—or even understood how a natural grassland ecosystem worked.

For scientists wanting to study the tapestry of life in an unspoiled natural grassland, East Africa's Serengeti savanna is a treasure. The entire ecosystem encompasses more than 9,000 square miles in northern Tanzania and southern Kenya, just east of Lake Victoria. The Serengeti shelters a paradise of animals, including more than 60 species of herbivores, carnivores, and primates, some 450 species of birds, and countless invertebrates. Nearly a fourth of Africa's estimated ten million big wild mammals live in the Serengeti.

Plains of high grass and shortgrass carpet the heart of the region, surrounded by woodlands and bush to the north and west. Southeast of the central plain, the land rises to the 10,000-foot heights of the Crater Highlands, a chain of extinct volcanoes whose ancient lava and ashfall created the savanna's fertile soil. Almost all of the Serengeti ecosystem lies within the boundaries of the Serengeti National Park and adjacent reserves. It is this protection that has made the area such a precious resource to scientists unraveling the secrets of life in the grasslands.

"I go to the Serengeti because it's the last gasp of undisturbed grassland in the world," says Dr. Samuel McNaughton of Syracuse University. "It's the prototypical grassland ecosystem, with plants and animals that have been evolving together for millions of years. They're all still there—and that's magical."

The day I arrived in the Serengeti, I met Dr. Markus Borner. He and his wife, Monica, are codirectors of the Frankfurt Zoological Society's Tanzania Wildlife Conservation Project. Based at Seronera, the headquarters of the park, they make aerial surveys to count animals, to watch for poachers and for settlers encroaching on parkland, and to track the movements of the wildebeest—the largest herd of migrating animals in the world.

"The Serengeti is full of mysteries," Markus said, "and it is a laboratory for landmark studies of animals and grasses. The more we learn, the more complicated it gets. Everything is interrelated—every insect, every blade of grass. We have to be careful not to tamper too much with something we don't fully understand. Right now we have scientists studying everything from the role of fire in savanna woodlands to the antipredator behavior of Thomson's gazelles."

From Seronera, I left on safari to see the incredible variety of life that survives in a seemingly simple world of grass. My guide was Bill Baker, a wildlife artist from England. He knows the Serengeti well after nine years of taking wildlife photographs to use as reference works for his paintings. We mapped a journey that would take us on a great zigzagging sweep across the central plain and south to Olduvai Gorge, up into the Crater Highlands and down along the Great Rift Valley, through the eastern mountains and west toward Lake Victoria, then north to the woodlands near the Kenya border.

During the first leg of our trip, we traveled against the tide of migrating wildebeest. It was dry season, and the rains that fall in East Africa from March to May had long ended. The wildebeest had grazed most of the green plain to stubble and dust and had set off on their annual migration to the west and north in search of fresh grass. They travel up to southern Kenya, to the Masai Mara Game Reserve that marks the northern boundary of the Serengeti ecosystem. They graze the high grasses there from August on, until the short rains of November and December lure them back south to the lush central plain of the Serengeti.

Wildebeest anticipate the rains, migrating to places where rainfall will soon produce a flush of green grass. They can detect rain as far away as 30 miles. "They see the lightning and they go," Bill told me. "They seem to literally follow the storms around the Serengeti." The annual migration routes of the 1.4 million wildebeest define the limits of the Serengeti ecosystem, and they are the most abundant big-game staple in the savanna food chain. Lions, cheetahs, and hyenas devour most of the animal, leaving the remains to other creatures. The last traces of the scavenged animal are its skull and horns, on which tiny moth larvae feed until they mature and can fly.

"It's nice to think of those moths finishing off the horn and flying up in a white cloud into the African night, like angels carrying away the last of the wildebeest soul," Bill said.

Everywhere we traveled in the Serengeti, the wildebeest had either just left, were about to arrive, or were there in noisy multitudes, darkening the gold savanna. At one point, Bill and I came upon thousands of the animals running in skeins toward Kenya—the first wave of the northbound migration. That evening in camp, I watched the sun set beyond a moving horizon of wildebeest silhouettes, galloping forward and kicking crazily against the hot orange sky. From afar, their deep grating honks sounded like the basso profundo chorus of millions of bullfrogs, and I heard them all night long.

It seemed that every animal species sent a representative to investigate our camp that night. I heard hyenas and jackals and hyraxes. A lion coughed. When a strange rustling grew closer, I looked out of the tent to see the eyes and horns of a buffalo, trying to hide its huge bulk in the bushes. Later on, a lone giraffe strolled around our truck, then gracefully bolted from my flashlight beam.

The eerie night spyings seemed an imagined fear in the cheery brightness of daylight, but tracks in the dust around the truck betrayed our many visitors. In a land so open and exposed, far more animals knew exactly where I was than I could see or sense. It was always dangerous to wander away from the truck; the shape of a person could be a profile of prey to a lion or a leopard. "If you walk out there, you're just another animal on the plains," Bill cautioned.

Southeast of Seronera on the shortgrass plain, we stopped in the center of a wraparound frieze of thousands of wildebeest, zebras, and gazelles. They were calmly grazing—the eternal scene of the African savanna. The wildebeest, each standing in a separate sphere of grass, seemed motionless, like statuary of the plains. Zebras mingled with the wildebeest. From a distance, their stripes wavered in the heat that rose in swirls above the grass.

No two zebras are striped alike, and nobody is certain of the purpose of their stripes. Despite their boldness, stripes may be a form of camouflage, some scientists believe. In the shimmering heat, the zebras that I saw seemed to blur together, and when they ran as a group, they merged into long rippling bands of black and white. In the grainy gray light of dawn and dusk, the times of the day when many predators hunt, the most vividly patterned animal on the plains can look almost invisible.

I had come in the dry season, when the Serengeti looks dusty and sere and incapable of nourishing the enormous numbers of animals grazing there. The existence of such a variety of wildlife in an environment so apparently austere is one of the great mysteries of grasslands. Studies in this region have revealed subtle relationships among the grazing animals. Zebras and wildebeest and gazelles may travel together for mutually beneficial reasons. Zebras can digest coarser tall grasses that other animals cannot eat. Wildebeest prefer short-to-medium-height grasses—and tallgrass cropped by the zebras. Gazelles favor the shortgrass areas where grazing by wildebeest has exposed low-lying, tender herbs. Instead of competing for pasture, these animals facilitate one another's survival.

Across the plain, away from the wildebeest, Bill and I navigated a course by compass, as if we were out on the open sea, to find the isolated spot where Clare Fitzgibbon, a Cambridge University scientist, was following a pack of wild dogs as part of her study of the antipredator tactics of Thomson's gazelles. Wild dogs are a rare sight in Africa. Most have been shot as vermin. Even in the Serengeti, where perhaps two hundred live, they are elusive. Only when they have pups do they remain in one place—and then just for about three months, until the pups learn to run with the pack. When we found her, Clare had located a small warren of tunnels and holes, where seven adults were caring for eight pups.

"They're so easy to watch right now," Clare said. "When the pups are small, the adults feed them here at the den, and they hunt twice a day, at dawn and dusk, like clockwork."

The pups huddled by the den, a clump of black noses, watchful eyes, and big Mickey Mouse ears. The adults were gamboling around in circles, sniffing and greeting one another, making high-pitched, twittering *(Continued on page 136)*

*I*n columns miles long, wildebeest follow rains toward greener pastures. These grass eaters, which number 1.4 million, join zebras, gazelles, and other grazers on a yearly 125-mile migration between northwestern and southeastern reaches of the Serengeti. Predators stalk the herd, picking off stragglers; a lioness (opposite) grasps a 350-pound wildebeest in a lethal embrace.

PAGES 132-133: Buffaloes graze in the Serengeti, a mesmerizing world of high grasses, shortgrasses, and woodlands that spread across 9,000 square miles.

G*aping yawn exposes huge canine teeth of a hippopotamus in Tanzania's Lake Manyara National Park. The sleepy gesture does express boredom at times, but when needed it serves as a warning to intruders to keep out of the animal's territory. Third largest land animal after the elephant and rhinoceros, the full-grown hippo stands 6 feet tall at the shoulders, has a girth of nearly 15 feet, and weighs up to 4 tons. It spends most of its time lolling in water and wallowing in mud baths, venturing ashore at night to feed. Surprisingly nimble on land, the hippo can run 30 miles an hour in short bursts. A voracious eater, the "river horse" neatly clips grassland vegetation with $2^1/_2$-foot-wide lips, consuming more than a hundred pounds a day.*

sounds, working each other up for the hunt. Less than half a mile away, a flock of gazelles grazed.

The dogs trotted back and forth, restless. Then, at some signal known only to them, they all began to stalk—fast. The dogs fanned out and broke into a run. The scattered gazelles lifted their heads in a ripple of alarm from near to far and took off racing and stotting in long, four-legged bounds. The dogs ran hard, like dark bullets. Closing in from three sides, they separated a mother and fawn from the flock. Singling out the fawn, they chased it down and flipped it in the grass; within minutes only skin and bones remained. The wild dogs returned to their den and fed their pups. After a while, the adults settled down to rest until dusk, when hunger would rouse them to hunt again.

To survive in a land with few places to hide and where their prey is faster than they are, the wild dogs use teamwork, one of many forms of social behavior fostered by an open-country environment. Socializing is a strategy common to grassland creatures, and scientists gain insight into the development of early man's group behavior by studying the other social mammals of the African grasslands they all once shared.

Of all the world's cats, lions are the only ones that socialize, and one morning Bill and I left Seronera to search for them with Craig Packer, a research scientist from the University of Minnesota. Craig carried a stack of lion "flash cards," displaying sketches of every lion in the Seronera area and the whisker spots unique to each individual. Lions' whiskers are much like human fingerprints; no two sets are exactly alike.

"The main thing I've been interested in lately is why are lions the only social cats," Craig said. "Very few animal species show the kind of coordinated behavior that lions have, and I'm interested in the evolution of that behavior."

We rounded a bend in the road and sighted 18 lions. "They're slinking," said Bill. "Aha, they see something," Craig said. "There's a female buffalo by the river—and a whole herd behind it." We edged closer to the river, and so did all 18 lions. One lioness took the lead and suddenly sprang forward, leaped on the buffalo's back, and brought her down. A second lioness moved in. "They've got her," said Craig. "Oh no, wait—here comes the herd to the rescue."

The buffalo herd approached the river; two bulls came forward, and one shoved the lioness off the downed buffalo's back. All 18 lions eased away in the grass, heads low in defeat.

"Actually, one of the primary causes of adult lion mortality is being gored by buffaloes defending themselves," Craig said. "It's quite risky to be a lion. Surprisingly, the advantages of their cooperative hunting are small. Lions are not the most efficient hunters, and when they are successful, they have to share the meal. There are advantages to being in a large pride, especially for lionesses protecting their cubs, but we can't explain the advantages in terms of cooperative hunting."

"We stand to learn a lot more in the Serengeti," he added. "But there are so many pressures on the park. There has been tremendous population growth around this region, and some of the land—which is outside the park but still within the Serengeti ecosystem—may be lost to the wild animals. In so many of the other grasslands around the world, we have no chance to know how the animals evolved with their habitat, and that's why it's so important to keep this place intact for a long, long time."

After Bill and I left Craig at Seronera, we headed for Olduvai Gorge, in the Ngorongoro Conservation Area southeast of Serengeti National Park. We set up camp at Olduvai, where gradual erosion of the grassy plain has exposed fossilized clues to the fine-tuned relationship between grasslands and their native wildlife.

"Ungulates and grasses originated simultaneously and began to flourish some 50 million years ago," says ecologist McNaughton. "They underwent changes together. As grasses developed more and more abrasive silica in their tissues in defense against the grazers, the animals developed high-crowned teeth that enabled them to break down the tougher grass fibers. The trend continued, and fossil records at Olduvai Gorge clearly show this remarkable progression. The grasses and grazers actually coevolved," he explains, "and the end result is a dynamic relationship between plants and animals in natural grasslands that has reached a balance after millions of years."

More proof of this harmony came in the 1960s, when Dr. Tony Sinclair of the University of British Columbia made a startling discovery about the wildebeest in the Serengeti. A plague of rinderpest—a deadly bovine disease—had cut the wildebeest numbers to 250,000. Eventually the disease was arrested, and the wildebeest population rebounded. At the same time, overgrazing by cattle was destroying the grasslands outside the park boundaries, while inside the park the grasses thrived even as the wildebeest numbers continued to increase. What kept the wildebeest from overgrazing?

Dr. Sinclair found that the animals were regulated by their food supply in such a way that their population would never exceed the capacity of the grasslands to feed them all. Annually about 5 to 10 percent of wildebeest become undernourished and die in the dry season, when the grasses are less nutritious. The loss in numbers keeps the wildebeest population compatible with the changing fertility of the grasslands, year after year.

In Africa, I had come to think of animals and grasslands in the same breath, and when I later visited the remnants of the North American prairie, I was struck by the paucity of wildlife. The prairie seemed forlorn without great herds of wild animals. As many as 60 million bison and millions of pronghorns once packed the North American prairie, which rolls across the central third of the continent—the richest grassland in the world. In Eurasia, uncounted millions of saiga antelope migrated seasonally across the harsh steppes—earth's largest grassland area and the one with the most severe climatic extremes. Like wildebeest, saigas can sense approaching weather. But instead of chasing rains, they flee the steppes' deadly blizzards, which can drive temperatures to minus 40°F.

Neither the saigas nor the bison could escape the hunter's gun, however. In the 1800s, these great herds—the largest masses of animals in history—were ruthlessly slaughtered, Eurasia's saigas by Europeans for skins and for meat, North America's bison by white settlers for robes and for sport, as well as for meat. Also, buffalo hunters became central figures in a U.S. Army ploy to win the West by eliminating the major food source of the Indians.

"Once the bison were gone, the prairie changed," says Dr. McNaughton. "By virtue of their enormous numbers, the bison maintained shortgrass prairies by trampling and grazing. Without the bison, the short, *(Continued on page 144)*

*S*urveying its territory, a male topi perches atop a termite mound in southwestern Kenya. Fond of high, green grasses, these hump-shouldered antelopes avoid the shorter plants, leaving them to such grassland species as gazelles. Above, a female cinnamon-chested bee eater deftly fields an insect. During courtship the male catches a flying bee or wasp, rubs out its stinger on a branch, then flicks the morsel to his intended.

FOLLOWING PAGES: World's tallest animal, a giraffe towers above gazelles and other creatures in an eternal African tableau, dusk on the Serengeti.

*O*nly its eyes and nostrils visible, an Orinoco crocodile waits for prey in a duckweed-choked pond on Venezuela's llanos. Slow to drain because of their base of thick clay soil, the prairie-like grasslands flood every May as rains overflow the Orinoco River and its tributaries. Animals of the region, such as the hoatzin (opposite), show deft adaptations to the annual inundation. A hoatzin chick dives into the water when alarmed. The young bird can swim underwater like a seal. The capybara (far right), largest of all rodents, grows to 100 pounds or more on a diet of water plants and grasses. Webbed feet help make it a good swimmer. With eyes, ears, and nostrils located on the top of its head, the capybara can monitor its surroundings while most of its body remains submerged.

palatable grasses gave way to taller, ranker grasses. It was once part of our common folk memory that the prairie looked different when the bison were there."

At the Samuel H. Ordway Jr. Memorial Prairie, in South Dakota, a visitor can find a haunting glimpse of the early day plains. I wandered out alone on the prairie, an 8,000-acre preserve of mixed grass owned and managed by the Nature Conservancy. The wind swept the treeless tract in long, moaning gusts. In its melancholy beauty the land carried its ageless feeling—a sense of freedom and limitless possibility borne by endless grass rolling to every horizon. On a distant hill, a few of the preserve's bison stood—shaggy silhouettes, silent as Indians, like ghosts of the grasslands.

Inspired by research done in the Serengeti, research scientist Glenn Plumb is comparing the grazing ecology of bison and cattle at Ordway. One afternoon, I joined Glenn as he painstakingly clipped plants from sample areas. All around us curving hills of grasses rose and fell like ocean swells.

"When you think of the Great Plains dominated by billions and billions of these plants, it is wondrous," Glenn told me. "A huge sea of these grasses that have evolved in a way that nothing else in the world has evolved—it's a beautiful sort of mystery."

The tallgrass, shortgrass, and mixed prairies of North America once covered an area 2,000 miles long and 1,000 miles wide from the Canadian north country to Mexico. Most of that prairie is farmed and grazed now. I came to appreciate how little of our native North American prairie remains when I visited several remnants of tallgrass prairie in western Minnesota. Geoff Barnard of the Nature Conservancy drove me past gingham-checked squares of Minnesota farmland to a scattering of prairie preserves—patches of tallgrass spangled with late summer's purple and yellow blossoms—tucked amid fields of wheat and sunflowers.

"Our prairie survives in islands now," Geoff explained. "And there are so many questions about the right populations of critters in these prairie islands. How many should there be? In isolation will they evolve into different species? Will we lose species? For every species we know in detail, there are dozens we know nothing about. We can't return the prairie to the way it was before settlement. And we don't even know what it was like then—it disappeared too fast. We're just trying to protect the biological diversity of these prairie remnants, because we don't know what answers they might give us down the road."

Grasslands worldwide have suffered in the wake of progress. In South America, commercial flocks of sheep and herds of cattle and horses have drastically altered the fertile pampas, where prehistoric forms of camels, rhinos, elephants, and bears once roamed. Ranching in Australia has consumed most of the continent's natural grasslands. Nevertheless, some 50 species of kangaroos and other marsupials thrive in a niche filled by wild, hoofed mammals in other grasslands. Kangaroos have one thing in common with Eurasia's saigas: Both can survive on sparse, poor quality vegetation that few other big animals, wild or domestic, can eat, making productive a land that would otherwise be barren.

Breaking camp at Olduvai Gorge, Bill and I drove to Ngorongoro Crater, a volcano that had exploded and collapsed inward about three million years ago. Now inactive, it contains the world's largest unflooded caldera. The 100-square-mile crater floor, big enough to hold Washington, D.C., with plenty of room to spare, shelters a miniature Serengeti, a Noah's Ark of Pleistocene animals cupped in the palm of Africa. We followed narrow hairpin turns leading 2,000 feet down into Ngorongoro. Near the center of the crater lies a shallow soda lake, its surface bright pink with waves of flamingos. On the hard-baked flats near the lake, we found a scattered group of rhinos standing still as boulders.

Unlike most herbivores of the savanna, rhinos are rarely preyed upon, except by man. Sadly, rhinos are all but gone in the Serengeti. They have been poached to near extinction for their horns. Ultimately the horns go to buyers seeking dagger handles or who believe that the horns are an aphrodisiac. We paused near a pair of rhinos. They were so perfectly motionless that it hadn't occurred to me that we had crossed the threshold of threat. Without warning, one charged—and stopped just short of our truck, huffing. Faced with a creature that—with a few swift jabs of its horn—could have wrecked the truck and killed us, I felt the odd chill of mortality.

We left quickly, humbled and relieved, and drove north along Africa's Rift Valley to the little-known region around Ol Doinyo Lengai, a black volcano that rises above a land the color of lion. Broad expanses of fine-spun grasses swept around the base of Lengai. A family of shy oryx with lance-long horns—said to be the unicorns of myth—grazed in the distance. All was golden and rustling and moving; sheets of wind whipped the tall-grasses. The land appeared so fresh and newly formed that to me it seemed a vision of the place where grasslands were born.

When grassy plains arose around the world, sometime after the age of dinosaurs ended 65 million years ago, geological events aided their success. It was an era of great continental uplifting, and the climate grew cooler and drier. Over the last several million years glaciers, advancing and retreating in northern temperate regions, opened huge areas for grasses to colonize. In the tropical belt, forests shrank back from the drier lands, and the adaptable grasses moved in.

Grasslands offered wide corridors for the migration of animals across continents. New land bridges that formed during glacial periods made it possible for some animals to travel between continents. Ancestors of the modern horse, for example, left North America, bound for the grasslands of Eurasia.

When the landmasses of present-day North and South America joined some two million years ago, many grassland creatures of the southern continent were pushed out by migrants from the north. Like Australia, South America had large populations of grassland marsupials, but the more evolved placental mammals from North America—deer, foxes, rabbits—ousted them from their niches. Millions of years ago, before a rift defined the huge areas that we now call Africa and Eurasia, a number of hoofed plains animals, as well as elephants and aardvarks, migrated north to Eurasia, while horses crossed down into Africa.

Like the animals, early men traveled the broad pathways of grasslands, and beginning with *Homo erectus,* they harnessed a force that still shapes the character of all grasslands: fire. Natural fires ignited by lightning have always swept across grassy plains, and fire actually maintains many such regions by releasing nutrients from plant litter, by clearing debris, and by exposing the soil and new shoots to sunlight. Once early man learned to control fire, he had the power to alter the composition of grasslands. Some experts say that Eurasia's steppes were once partly forested, and that clearing and burning by people long ago turned the woodlands into grasslands.

While fires occurring in August through October (the late dry season) are destructive to the Serengeti's grasses and woodlands, burns in June and July (when moisture helps protect the base of grasses) consume the heavy mat of dead grass and stimulate a new flush of green growth. Late season fires are especially troublesome in the northern Serengeti, a favorite haunt of poachers. Well-armed elephant hunters, as well as local tribal people stalking game for food with poison arrows, set fires to cover their tracks. To confine the damage of those burns and to study the effects of fire on the ecosystem, Neil Stronach of Cambridge University creates firebreaks and burns patches in the north.

"Without these early season fires, the whole northern Serengeti would burn later on," Neil explained. "It doesn't matter so much that you have fire here, but when it's so frequent and when it's caused by humans, it's not a natural thing any more. Such fire is changing the trees and grasses of this savanna woodland, and we're trying to protect what belongs here naturally. Once you get species dropping out, you might not get them back again."

Neil stopped and showed Bill and me the spiral awns of red oat grass. A seed clung to the tip of each slender awn, which resembled a miniature devil's tail. "This grass is promoted by fire," Neil explained. "The awn straightens and curls in response to humidity and dryness, digging its way into the ground and planting the seed. Then the heat of fire stimulates the seed to germinate. You do see some lovely things that have evolved with fire."

We burned black patches along a 30-mile-long swath just south of the Kenya border. Around us were the migrating wildebeest herds. The savanna where the animals had grazed was clean-shaven. "This place is like a football field," Neil said. "I like to see the wildebeest. I know their grazing will make a nice firebreak for me." That night we set up camp beside the Mara River. Upstream, hippos grunted and splashed, and when we played our flashlight beams across the surface of the river, the eyes of crocodiles gleamed. When we flicked off the lights, the black water twinkled with the reflections of stars. "The Serengeti would be unique even if there had never been men on this continent, changing the land," said Neil. "In all of Africa, in all time, there has never been a place like Serengeti."

Soon our trip would end, and I wanted to get a closeup look at the great beast that roams Africa's grasslands and woodlands—the elephant. In the Serengeti, elephants have been heavily poached for their ivory tusks, and like the rhinos, they appear less often now. In Kenya's Amboseli National Park, east of the Serengeti, the elephants seemed relaxed, and we approached them with ease as they dusted themselves, munched on acacias, bathed in a swamp, and nursed their young. I had to remind myself that they were dangerous and could trample or gore me in an instant, if provoked.

On our last morning in Amboseli, Bill and I searched in vain for a sack of fruit we had left inside the truck the night before. Then Bill spotted fresh elephant tracks in the dust around the truck, and we reconstructed what had happened. An elephant had sneaked up in the night, reached its trunk inside the truck's open back, and snitched our bag of food. "Only an elephant could have taken the food so quietly," said Bill. "And they love fruit." Not far from our camp we came across a bag, like the one that had held our food, snagged in an acacia. Farther on, we came upon 15 elephants. They eyed us. One female was tuskless, and Bill recalled that a tuskless female had been known to raid camps around Amboseli. I stared at her wide-set, knowing eyes and pictured her as the guilty one. They all turned away and ambled off across the grass, trunks twirling, ears flapping, tails swinging, and, I imagined, gloating over having gotten the best of two humans.

What do these animals, which seem so clever, actually know? They know that parks mean safety, and they retreat to them when pursued. Some poached herds increase their nighttime activity to avoid man, and the memory of that behavior persists in their descendants. These dramatic findings come from the work of Iain and Oria Douglas-Hamilton in Tanzania's Lake Manyara National Park, southeast of Ngorongoro. Their studies reveal strong kinship ties among elephant family units, which may remain stable for more than a century. Modern elephants appeared some four to five million years ago, about the same time as the early ancestors of man. Could the behavior of elephants hold clues to the development of human society?

Contrary to myth, no elephant graveyards exist. But the humanlike reactions of elephants to the death of their own kind exceed fictional tales of the graveyards. In Uganda in the mid-1960s many elephants were killed, and their ears and feet were put in a shed for later sale. In the night, a group of elephants crashed into the shed and took the ears and feet away.

Elephants have buried the bones of their kin and the bodies of people they have killed. Tusks have been found smashed against rocks, presumably by elephants. They have been seen trying to pull tusks from a carcass, as if to keep the ivory away from the men who had shot the elephant.

What strange communication passes among elephants? Katy Payne, a research scientist with Cornell University, may soon have some answers. In 1984, Katy spent some time observing elephants at the Washington Park Zoo in Portland, Oregon. "Every now and then I felt a throbbing in the air," she told me, "like what you feel when the low notes of a church organ are being played, or when you feel thunder. I wondered whether I was feeling sounds that were too low for me to hear. I recorded the vibrations with a tape recorder that picks up frequencies below the range of human hearing. When played back faster, the very low frequencies became audible. The sounds were something like the moos of cows, and I could hear separate voices answering each other. I realized that elephants might have a tremendous communication system that no one knew about."

Katy traveled to Kenya to record the elephants of Amboseli with researcher Dr. Joyce Poole. "My research is in a very early stage," Katy said, "and this is a difficult thing to study. We don't know what the elephants are communicating. Unlike high sounds, very low frequencies don't dissipate over great distances, so these elephant sounds are traveling far. Since we can't detect their low sounds with our human ears, it's like discovering a secret channel for another species."

"This could be just the beginning," Katy added. "Who knows what other animals are communicating in ways we haven't even considered? It's so important that we preserve the great African grasslands with all their wildlife, or we might lose an extraordinary chance to understand what those animals really are."

And, perhaps, to learn what we really are.

JIM BRANDENBURG (BOTH AND 150-151)

*N*orth America's fastest wild land animals, pronghorns bound across the Samuel A. Ordway Jr. Memorial Prairie, a preserve of virgin grassland in South Dakota. For thousands of years pronghorns and vast herds of bison shared America's prairies. Two stolid bison (opposite) face into a Minnesota blizzard. Their head-on position to the storm keeps the icy wind from blowing under their fur and freezing them to death. Overhunting nearly exterminated bison in the United States; in 1900 only 39 remained in the wild. Today some 55,000 roam ranches and preserves.

FOLLOWING PAGES: Goldenrod proclaims autumn on South Dakota's prairie. These tallgrasses—big bluestem and sand reed—once carpeted the plains of the Midwest.

G*ayfeather blossoms spike a tallgrass prairie in Wisconsin. A male Wilson's phalarope (below) tugs grass to conceal its nest. The larger, more colorful female Wilson's courts the male near prairie ponds. After laying eggs, she abandons the nest, leaving the male with the job of hatching and raising the chicks.*

*P*rairie wildlife of South Dakota: A newborn pronghorn crouches motionless amid shortgrass; in about a week after its birth the fawn will run with its mother. Females graze a short distance away from their offspring, careful not to reveal their newborn's hiding place. Opposite: Darker feathers have begun sprouting on four young ferruginous hawks. When fully grown, the hawks will have wingspans of well over four feet, and some observers will mistake them for eagles. For prey, the hawks favor prairie dogs, such as the two nuzzling in an act of recognition (far right, lower). Before croplands replaced natural prairie, these gregarious rodents probably ate more grass than bison and pronghorns combined. A Texas prairie dog town discovered at the turn of the century spread across 25,000 square miles and housed an estimated 400 million animals.

FOLLOWING PAGES: A stream shields a
pasqueflower from flames creeping
across a Minnesota preserve. Wildfires
ignited by natural causes, such as
lightning, have for millions of years
stimulated the growth of grass by
clearing away accumulated thatch,
killing woody plants, and hastening the
return of minerals to the soil. Green
blades of fresh prairie grasses quickly
sprout from undamaged turf below the
scorched surface. Without regular
burning, some of the world's grasslands
might have disappeared long ago,
overgrown by forest.

JIM BRANDENBURG (ALL AND 156-157)

155

Arid Realms

The Deserts

By Thomas O'Neill

Flowering tendrils garland barrel cactuses in the Sonoran Desert, which sprawls across corners of California and Arizona and extends into Mexico. The region supports more than 300 species of cactuses, including the pole-shaped cardon marching up the slopes at far left. Desert plants divide into two main groups, those that resist excessive dryness and those that evade it. With few exceptions, less than ten inches of rain fall annually on the world's arid lands.

FOLLOWING PAGES: Furry spines diffuse hot sunlight and help keep teddy bear cactuses cool on a rocky Sonoran slope in Arizona.

TOM BEAN (LEFT); N.G.S. PHOTOGRAPHER BATES LITTLEHALES (160-161)

159

JOHN GERLACH/TOM STACK & ASSOCIATES (BELOW); N.G.S. PHOTOGRAPHER BATES LITTLEHALES (RIGHT, UPPER); DAVID MILLERT/TOM STACK & ASSOCIATES (RIGHT, LOWER)

*S*kyscrapers of the Sonoran,
saguaros reach heights of 50
feet. The Gila woodpecker
(upper) escapes the desert
sun by chiseling nest cavities
into the cactus. In summer,
the waxy white blossoms
(above) of the saguaro give
way to sweet fruits with
juicy, watermelon red pulp.
From the fruit, Papago
Indians make such products
as flour, jam, and wine.

The Deserts

*H*iya! *Hiya!* His cry as insistent as a raven's, the veiled man urged us to hurry. He looked imposing standing on a shattered boulder, his blue robe swelling in the desert wind. Only his dark nose and eyes showed from the cloth wrapped around his head. He was a Tuareg, member of a once nomadic people who steered as well as raided caravans that crossed the Sahara. Now he was left to guide visitors into the Tuareg's former sanctuary, the Tassili-n-Ajjer, a foreboding mass of mountains in the heart of the desert.

The Tuareg pointed to his watch, the only clue that he belonged to the 20th century. His ancestors would simply have gestured at the sun, now starting to mount rapidly in the sky, its yellow color burning to white. Soon the torrid heat would become the enemy, catching us defenseless on the steep mountain pass. Already Jean Claude, a young Frenchman from Brittany, had collapsed from heat exhaustion. A friend poured water on his pale, frightened face. Precious water that was. None was to be had atop the Jabbaren plateau, our destination.

Reviving Jean, we pushed on and finally reached the top of the plateau, 5,000 feet above the sandy plain in southern Algeria. Keep out, said the glaring sun, the naked rocks, the parched earth untouched by rain for more than a decade. Soon the Tuareg guide called a halt, motioning us into the shade of a low, overhanging cliff. At last, here was what we had come for.

I saw on the smooth rock face in front of me the figure of a woman with a basket balanced on her head. Nearby, three leaping archers chased an antelope; farther on, men in loincloths watched over a herd of long-horned cattle. At the base of another cliff we were stopped by the sight of a fading, but still vibrant painting of a procession of dancing women; their fluid, abstract shapes could have been painted by Matisse. On another rock canvas appeared the unmistakable spotted face and lanky neck of a giraffe.

Giraffes, cattle herds, and dancing women in the middle of the Sahara? Who painted these strange frescoes and why?

Hard as it is to imagine, the Sahara—world's largest desert and one of earth's most inhospitable regions—was once a lush, river-laced expanse, home to prehistoric cultures. When the Ice Age gripped Europe, a Mediterranean-type climate prevailed in North Africa. Giraffes and elephants roamed savannas similar to today's Serengeti Plain in East Africa. Lime trees and cypresses soughed in the mild breeze. Against this temperate backdrop appeared dark-skinned peoples in the Tassili mountains. Armed with clubs and bows and arrows, they tracked wild game. Later other people moved in and raised huge numbers of cattle, all the while living in the natural shelters of the cliff overhangs.

Their achievements might have been forgotten utterly had not these ancient peoples taken ocher pigments and produced paintings of remarkable skill and expressiveness on the walls of their homes. As it was, the extent of these Saharan cultures remained unknown until a French army officer on patrol in 1909 noticed images of animals on the walls of wadis, or gullies. Since then, thousands of rock paintings and a score of engravings have been discovered in the Tassili highlands, now an Algerian national park.

Archaeologist Henri Lhote, one of the first to study the area and copy its frescoes, called the Tassili "the greatest museum of prehistoric art in the world." He speculates that the early inhabitants of the Sahara used their paintings in rituals related to sacred cults, or as talismans to bring luck during a hunt, or, in the case of the frequent cattle paintings, as a means simply to inventory property. Paintings first appeared around 9000 B.C., in the late Stone Age, experts estimate, and continued to the beginning of the Christian era. Around 1000 B.C., during the middle period, the rock paintings and engravings depict horse-drawn chariots. In their final phase the prehistoric artists frequently drew camel caravans, an unmistakable sign that the climate was changing. Soon the herdsmen would abandon the increasingly arid Sahara, leaving it to its enormous silence and to the occasional visitor who might dream of when the Sahara was green.

As a visitor, I had come to this region to experience a desert at its mightiest—to feel my throat tighten and my face flush in heat that pours in white light from the sky, to submit to an emptiness so encompassing that I could travel for days and, with the exception of a cargo truck or military vehicle, see absolutely no sign of life—animal, plant, or human. I discovered the exhilaration of scaling a sand dune hundreds of feet high, its grains as orange as a robin's breast. I feasted my eyes on the luxuriant green of palm groves in remote oases. I sweated with anxiety as a dust storm swallowed the landscape in a yellow fog and stung my face and eyes. From a perch on a mountaintop at sunset I let silence and solitude wash over and cleanse me like an ocean wave. No other environment that I have known can dominate one physically and mentally like the desert, and yet no other place confers such an imminent sense of adventure, or leaves one with such a deep appreciation of the stubbornness of life.

Deserts are defined as arid and semiarid environments. They cover a third of the earth's landmass, penetrating every continent. They range from the enormous—the Sahara covers an area almost the size of the continental U.S.—to the merely large. The Namib, one of the smaller deserts, extends 1,200 miles down Africa's southwest coast, about the distance from New York to Miami. Deserts exist mainly in an exact configuration, girdling the globe around the Tropics of Capricorn and Cancer, zones of hot air that have shed their moisture in the tropics. The band of aridity in the Northern Hemisphere runs from the Sahara through the Arabian Peninsula and into Iran and India, as well as across the American Southwest and Mexico. Below the Equator, the ring of arid lands encompasses the outback of Australia, the Namib and Kalahari Deserts of southern Africa, and the Atacama and Peruvian Deserts on the west coast of South America.

Rain-poor regions that fall outside the two great aridity corridors are often rain shadow deserts. They owe their existence to mountain ranges that rise in the vicinity. Warm, damp air approaching a wall of peaks is deflected upward by the barrier and cooled, so that clouds form and rain pelts the windward slopes. By the time the air currents cross the mountains and descend on the leeward side, they are drained of moisture, leaving deserts in the dry shadow of the rainy slopes. One such area, the Patagonian Desert, lies in the lee of South America's Andes.

The Sierra Nevada in North America dries out parts of California and Nevada. In some of these areas rain forests with towering fir trees crowd the western slope, while on the eastern side lizards dart among clumps of sagebrush. Asia's upland deserts also lie outside the global belt of dry zones. Aridity dominates, however, not because of mountains, but because of the deserts' location in the middle of a huge continental landmass. Sweeping inland for great distances, air currents lose most of their moisture, and by the time the wind reaches the Asian interior it blows dry across the parched land.

A region qualifies as arid if it receives less than ten inches of precipitation a year. Some deserts are bone dry. Chile's Atacama Desert annually averages in places only 4/100 of an inch of moisture, mostly fog. On the other hand, in southern Arizona some 18 inches of rain fell in 1984 on parts of the Sonoran Desert, causing saguaro cactuses to waterlog and split open. The world's most improbable desert is icebound Antarctica. In this so-called cold desert, a mere inch of precipitation—all snow—falls yearly on the raw polar interior.

Another measure of a desert, the aridity index calculates how much rainfall the sun's energy could evaporate in a year. In the eastern Sahara, the aridity index hits its peak of 200, meaning that the sun could soak up 200 times the mean annual precipitation. An inland sea 15 feet deep would dry up in a year.

By journeying to North Africa in April, I avoided the hellish heat of summer, when the mercury soars daily past the 100-degree mark. The highest temperature ever recorded—136°F—occurred in the Sahara at Azizia, Libya, in September 1922. As my two companions and I made our way from Tunisia into the stark desert wasteland of central Algeria, we never once saw a cloud—just a glaring, metallic blue sky. Daytime temperatures rarely wavered from the low 90s. We dubbed our trip "Shake and Bake" as we maneuvered our Land-Rover down endless tarmac roads and across the open desert on *pistes,* or tracks, marked, if at all, with stone cairns or automobile carcasses.

If we couldn't urinate three times a day, then we weren't drinking enough liquid, warned James Locke, the trip leader, who had gained his expertise serving in the British military in Libya and Saudi Arabia. Try gulping a half gallon of warm, sour, chemically purified water a day, though. It's hard, especially when I often did not perceive myself as sweating, the dry heat evaporating perspiration before I could even wipe my brow.

And yet, one can freeze to death in a desert. In China's windswept central desert—the Gobi, with an average elevation of 4,000 feet—the temperature stays below freezing for much of the year. On the Colorado Plateau in North America, a desert traveler may wake up under a blanket of snow in winter. Cold nights typify deserts. Sparse in vegetation, the bare desert surface retains heat poorly, and the heat escapes back into the air at night. With its low humidity, the air does not capture much warmth either. After the blazing Saharan days, the temperature at night would drop 50 degrees and force me into a down-filled sleeping bag.

Deserts often bring to mind vast sand seas where dune follows dune like frozen waves. Actually, sand dunes constitute less than a tenth of all desert surfaces. In the Sahara we saw by far more mountains than dunes—dark, low ranges of cracked rock, anvil-shaped outcrops enveloped in heat haze, volcanic plugs that rise from the ground like citadels. And then suddenly the land turns griddle-flat, and we come to a sterile gravel plain, or *reg,* that extends endlessly. British writer Jeremy Swift captures the terrain's character: "You travel a hundred miles and make your camp on a patch of gravel identical to the one you left the day before. At dusk you do not look for a good place to stop; you simply turn off the engine and let the car coast to a halt, and that place will be no better and no worse than any other for a hundred thousand square miles around."

The desert wears myriad faces—from the polychrome canyons and bizarre eroded pinnacles of the Colorado Plateau, to the rusty-red dunes that run for miles across Australia's Simpson Desert. In Ethiopia, white salt flats stipple the ground; in Mexico, black volcanic cones glower above forests of cactus.

For all their guises, the desert lands, I found, possess an essential uniformity, one of harshness and desolation. No traveler can help but feel vulnerable in an environment where a person without water can die in a matter of hours. A vehicle breakdown, the misreading of a map, a dried-up well, a sudden sandstorm: Each carries with it the potential for tragedy. On a grueling 500-mile drive, mostly across deep sand, between Tamanrasset, Algeria, and Agadez, Niger, James Locke once counted 157 abandoned vehicles. "Do you realize," he asked grimly, "that each vehicle represented a crossroads in someone's life?"

I knew exactly what he meant. Several years earlier, I had been driving a flatbed truck on a soft sand track in Australia's Tanami Desert. Suddenly I spotted a lizard in my path, a three-foot-long goanna. Instinctively I jammed on the brakes. They locked and, as if in slow motion, the truck slid and toppled over. After climbing out unhurt, I found a scrub tree nearby, and in its precious shade I sat down and waited for help. I tried to comfort myself with the knowledge that my accident had occurred on a well-traveled road. By desert standards, however, that could mean a vehicle every hour, at best. The sun seared the gaunt countryside around me. After half an hour I wanted help badly. Two very long hours passed before Aboriginals in a truck happened by and rescued me.

Still, for all its forbidding characteristics, the desert is legendary for exerting an almost mystical attraction. "The weird solitude, the great silence, the grim desolation are the very things with which every desert wanderer falls in love," wrote John Van Dyke, a devotee of arid lands as well as an American art historian.

The desert, however, contrasts its compelling bleakness with another kind of fascination: palm-fringed islands of life flourishing in the midst of aridity. Some 90 major inhabited oases dot the Sahara. They owe their existence to some of the world's largest underground supplies of water, which exist directly beneath this sunbaked wilderness. Dr. John Harshbarger, former head of the hydrology department at the University of Arizona, who has studied oases in western Egypt, explained: "The entire northern portion of Africa is underlain by huge aquifers of fossil water dating back as far as 2.5 million years. Rain that once fell more frequently in the region would run off mountains and soak through rock and soil, gathering into underground rivers that flowed into basins of impermeable rock."

One of the most extensive of these aquifers, the Savornin Sea—named after the French hydrogeologist who discovered it in 1947—covers an area as large as the state of Texas. Oases are created either when erosion of the desert surface exposes the water-bearing stratum, or when fracturing of the earth's crust tilts the rock layers close to the surface, making the water accessible by well.

"And the system is still being recharged." Dr. Harshbarger added. "Water from rivers and springs in the north and from lakes and savannas in the south in Sudan is running underground into the aquifers. In Egypt the aquifers are so full they're discharging into the Mediterranean."

One of the desert's delightful surprises happens on certain spring or summer days: Overnight a rain shower hits, and within days a drab khaki-colored waste has been transformed into a garden of riotous color. I caught the spring show one March in Anza-Borrego State Park in California's corner of the Sonoran Desert. I woke to classic desert weather—the heat dry and as crisp as starch, the sky lapis blue, the air shimmering on the horizon. I set off up Coyote Canyon, and colors began ambushing me from all around. Waxy, lime-colored blossoms sprang from the thorny limbs of cholla cactuses. Yellow flowers flashed like corsages from barrel cactuses. A fiery grove of tall ocotillos enclosed me, their skeletal limbs tipped in brilliant crimson. Purple sand verbena and white dune primroses fringed a sandy path. Tracking a red racer—a swift, red-tainted snake—I came upon one of the loveliest wildflowers of all, the desert lily, or ajo, with its tapered white cup striped in pewter.

That the desert can masquerade as a garden and bloom on short notice represents a triumph of adaptation. The desert lily, a perennial, may dry out and disappear aboveground, but as deep as 18 inches underneath, a bulb is storing nutrients. Given the right amount of precipitation, the plant will suddenly flower in all its glory, even after being dormant several years.

Many desert annuals—the desert dandelion, the dune primrose—survive with the aid of a growth inhibitor, a water-soluble chemical that coats the seed. Only rains of a quarter inch or more can dissolve the film and permit the seeds to germinate. "Otherwise, the flowers get faked out," says Mark Jorgensen, a park naturalist at Anza-Borrego. "If they get a heavy dew and the next day the humidity is 3 percent, without that seed coating you'll get a bunch of false starts."

Desert rainfall often erupts in the form of cloudbursts, when several inches can fall within a few minutes. Racing down the gullies and ravines of steep mountains, tearing across hard-baked ground with little vegetation to impede its course, the runoff can quickly turn into a flash flood. Such an onslaught of mud and rocks is powerful and deadly. Once, in the Mojave Desert, a flood picked up a locomotive, dragged it for a mile, and buried it in mud. Mark recalls a banker "who was driving across the park when a storm broke. He hit this flash flood unawares as it was crossing the road, or else he saw it and tried to go through it, but the water picked up his car, swept it over a cliff and down a canyon. He was thrown out and killed. Must have been a hell of a wall of water."

In California's Death Valley National Monument, a beavertail cactus unfolds bright blossoms. Death Valley, North America's lowest, driest, and hottest region, registers only 1.5 inches of rain annually. The scant moisture, however, encourages hardy plants such as the beavertail to flower. Seasonal showers bring forth splashes of color in February and March throughout the desert west.

*A*gainst the Ajo Mountains, organ pipe cactuses and brittlebushes cluster in Arizona's Organ Pipe Cactus National Monument. More than 500 plant species grow within this Sonoran preserve. All deserts pale when compared with the botanically diverse Sonoran. Like most cactuses, the organ pipe conserves moisture by manufacturing food inside its thick, insulated stem—a function of green leaves on plants in wetter climates. In times of extreme dryness the agave (lower) folds shut, shielding much of its fleshy surface from sunlight. Long spines of the hedgehog cactus (below) protect its blossoms and moist fruit from roving animals.

Driving out of Tamanrasset in southern Algeria, James Locke and I were hit by another of the desert's unnerving surprises, a dust storm. It turned day into hazy twilight. The air thickened; shepherds held cloths to their faces; donkeys brayed in distress. Either we sweltered in the closed cab, or we left the windows of the Land-Rover open and were lashed by a hot wind filled with stinging dust.

Desert winds go by many names—the *sirocco* in Algeria, the *khamsin* in Egypt, the *harmatten* in the southern Sahara, the *shamal* in Arabia, the *jeggos* in the American Southwest. Some of the Saharan storms fling so much material into the upper atmosphere that African dust ends up tinting sunsets in Miami and coating snowy peaks in the Alps. German naturalist Uwe George estimates that a million metric tons of sand and dust are blown out of the Sahara each day, enough to fill a freight train a hundred miles long.

At times the winds of sandstorms approach hurricane velocity, lifting grains of sand—heavier than dust—off the ground and transporting them in dense moving walls up to six feet high. A similar storm in 524 B.C. is mentioned by the ancient Greek historian Herodotus. He noted that a Persian army of 50,000 men, led by King Cambyses, vanished in a ferocious sandstorm in Egypt's Western Desert. The account so intrigued Gary Chafetz, a novelist from Boston, that he decided to mount an expedition to see if any traces of the lost army could be found 2,500 years later. Chafetz carefully studied maps, investigated the area, and talked with scientists and historians. Then in 1983, with supporting grants from Harvard University and the National Geographic Society among others, he began digging 75 miles southeast of the Sîwa Oasis, not far from the border of Egypt and Libya. The result: Some 400 to 500 Persian-style cairn graves were found. Selective excavations, however, uncovered no human bones or artifacts, leaving details of the fatal episode still shrouded in mystery.

Not the least of the many puzzles in the desert are the sights and sounds that play games with one's mind. On his journey across the Asian deserts to meet the Mongol ruler Kublai Khan, Marco Polo heard weird, ghostly noises. "Often you fancy you are listening to the strains of many instruments," he wrote, "especially drums and the clash of arms." What Marco Polo described was an unpredictable phenomenon known as "singing sands."

I never heard the sands perform, but the travelers that have liken the sounds that emanate from dunes to everything from a steamship siren and an airplane to the bellowing of cattle or the ringing of bells. Desert geologist Farouk El-Baz, formerly of the Smithsonian Institution, has heard sands sing in the Sinai and in China's Taklimakan, and he has taped them in California. "The only feasible

Playa puzzle: A 100-pound rock lies at the head of the trail it etched as it scooted across Racetrack, a dry lake bed in California's Death Valley National Monument. Scientists don't know what causes the area's rocks to move. Some speculate that they slide on gusty winter days when icy condensation glazes the playa. Others suggest that high winds spawned by thunderstorms may propel the rocks after rain softens the surface and makes it slick.

GEORG GERSTER

physical explanation for singing sands," Dr. El-Baz told me, "is that dry sand grains become electrically charged when they slide down a slope and are rubbed together. The noise comes from all the electrostatic charges given off. And when this effect is multiplied over and over, there can be a huge booming or high-pitched screeching, like a woman screaming. One night in the Sinai I heard such a chilling cry floating from the dunes. I can see why Arabs think that a *djinn,* a spirit of the dead, is calling."

Another desert deception I saw many times: a shimmering lake. The "lake" was a mirage, of course. A mirage forms when rays of sunlight are bent by a layer of warm air caught between layers of cooler air just above the horizon. When this happens a refracted mirror image sometimes hovers near the ground. That cool blue water that looks so inviting is only a reflection of the clear sky above. I cannot believe that anyone would be fooled for long by a mirage, despite the legends of caravans led to their doom by the blurry apparition of water. But the illusion is distinct enough and its appearance so frequent that I jokingly came to regard a mirage as the most common landform in the desert.

Ironically, for an environment whose very essence is defined by the scarcity of water, the desert wears on its face repeated evidence of seas, swamps, and downpours. In an abandoned rock quarry outside the oasis of Beni Abbes, I began picking out shapes and outlines in the rocks that proved to be fossil remains of shells and aquatic plants. I was treading on an ancient seabed.

In southern Morocco, a 300-foot-high ridge contains a petrified coral reef, built in a warm sea 350 million years ago. Near the oasis of Agadez in Niger, an eerie graveyard of hundreds of dinosaur skeletons has been found among sand dunes in what once was a tropical swamp. And on the banks of dried lakes in North Africa, bone harpoons litter the ground, evidence that humans survived as fishermen in the Sahara.

Even the great *ergs,* or sand seas, of the Sahara bear a waterlogged pedigree, I learned from Dr. El-Baz. "Many people," he said, "are taught that wind erosion makes sand, but we see today that this is nonsense. The rounding of sand grains couldn't happen in a windy environment. Grains colliding together would break and be angular. No, you find rounded grains at the mouths of rivers and at the bottom of lake beds."

What happened, Dr. El-Baz reasoned, was that during moist interludes in the Sahara millions of years ago, rivers carried a huge cargo of eroded material. The rushing waters pulverized, ground, and smoothed the debris as they flowed into the North African interior and emptied into natural basins to form shallow seas. Here vast quantities of sand accumulated. Only when the seas dried up did wind forcibly enter the picture and mold the sand into today's magnificent dunes.

In the first days of my Sahara odyssey, when the road skirted the Erg Oriental, or the Great Eastern Sand Sea, we had stopped to satisfy my desire to hike to the top of a dune. After scampering up the soft mountain, my footsteps filling in behind me, my breath racing, I gazed upon a panorama of beautiful curving shapes. In the outlines of the dunes I saw the sweep of scimitars, the slope of naked shoulders, the crest of pyramids, the repose of serpents.

The most desolate of all water stains on the desert has to be the salt flat, the evaporated remains of a lake bed. Perhaps the most infamous salt flat is found in Death Valley. There a 200-mile-long pan sinks 282 feet below sea level, the lowest point in the Western Hemisphere. Surface temperatures of 190°F, among the world's highest, have been recorded on this glaring depression, referred to by scientists as a "chemical desert." Nothing can live there—except bacteria.

Despite their forsaken aspect, salt flats have proved immensely valuable to humankind. Blocks of salt dug from sterile lake beds in North Africa formed the basis of the great trans-Saharan caravan trade during the Middle Ages. The societies of central Africa prized salt so highly as a dietary supplement (the traditional fare of boiled meat, grains, and vegetables provided almost no salt) that merchants would buy salt for its weight in gold. Today a few camel trains still transport salt from heat-blasted flats in Mali, Niger, and Ethiopia.

Of the huge inland seas that once flooded parts of present-day deserts, a few remnants still exist. One of them, the Dead Sea—which spills across disputed territory between Israel and Jordan—lies 1,312 feet below sea level. It is the lowest body of water on earth. In Utah the 2,000-square-mile Great Salt Lake is among the last traces of prehistoric Lake Bonneville. Distant centuries ago it covered an area about equal in size to Lake Michigan.

Nothing in the desert inspires me more than the plants and animals that can live in an environment made hostile by scarce water and abundant heat. I heard all about the heroics of flora in arid lands one March weekend when I drove through gaunt basin-and-range desert in southwestern Arizona near the Mexican border. My guide was Hal Coss, a converted desert rat from New Jersey. A former ranger at the Saguaro National Monument near Tucson, he loves the fact that the desert imposes tough demands on its living things; this way, he feels, the desert stays pure and uncompromising, a sanctuary for survivors.

As we plowed across a sandy wash, Hal pointed out a spray of mesquite trees on the bank. A mesquite taproot may extend more than a hundred feet down, he told me. Crossing the basin floor, we skirted creosote bushes. These Southwestern shrubs vigilantly guard their water rights. Through their roots they secrete a poison that kills off young creosote plants trying to claim their living space. Incredibly, one of these desert stalwarts, a creosote bush discovered in the Mojave Desert, dates back more than 7,000 years.

We avoided patches of fuzzy-looking teddy bear cholla, which isn't cuddly at all; its long barbs can impale birds and pierce the side of a leather boot. Without spines, though, most cactuses would not make it in the desert. A sheath of spines helps deflect the direct blast of sunlight, Hal told me. In addition, spines dissuade rabbits, wood rats, and other creatures from dining out on the plant's flesh. Moreover, spines have taken the place of leaves, enabling the hundreds of cactus species in North and South America to lower their water needs. Photosynthesis occurs instead in the trunk and limbs, whose thick, waxy skin further saves water by slowing evaporation.

Stands of creosote and hillsides of lemon-colored brittlebush frequently gave way to groves of the giant saguaro, symbol of the Southwest and among the tallest members of the cactus family. Saguaros can live up to 170 years and can grow as tall as a five-story building. More die from freezing or lightning strikes than from drought. Known as the desert's water towers, saguaros are renowned for their ability to hoard liquid. With its shallow, *(Continued on page 184)*

LOREN MCINTYRE (BOTH)

*W*aves of sand blown by prevailing southerly winds crest in this nearly mile-high dune on Peru's coast. The world's smallest and driest deserts, the Atacama and Peruvian hug a narrow stretch of shore along South America's Pacific coast. To the east, Andean peaks rob rain-bearing clouds of their moisture. To the west, cold ocean currents cause atmospheric inversions that inhibit precipitation. A layer of cold air underlying a layer of warm air spreads a cloak of fog (opposite). In some parts of these deserts rain may come only once in a lifetime.

FOLLOWING PAGES: Bushmen of Africa's Kalahari Desert fill ostrich eggshells from a shallow pond left by a rainstorm. Fewer than a thousand of these hunter-gatherers still wander the vast Kalahari, foraging for roots and fruits of succulent plants.

*S*olitary gemsbok in the Namib Desert leaves deep tracks that will disappear in the next scouring wind. Some of the world's highest dunes rise in the Namib, which blankets a slender area of Africa's west coast from Angola into South Africa. The gemsbok subsists on grasses, bulbs, tubers, and pockets of moisture. A gemsbok skull (opposite) shelters a small drift of sand on a wind-rippled dune.

PAGES 182-183: Flamingos gather at Sandwich Harbour. Along this coast, the Atlantic halts the march of dunes across the Namib. About every ten days fog rolls inland from the ocean, sustaining the desert's wildlife.

180

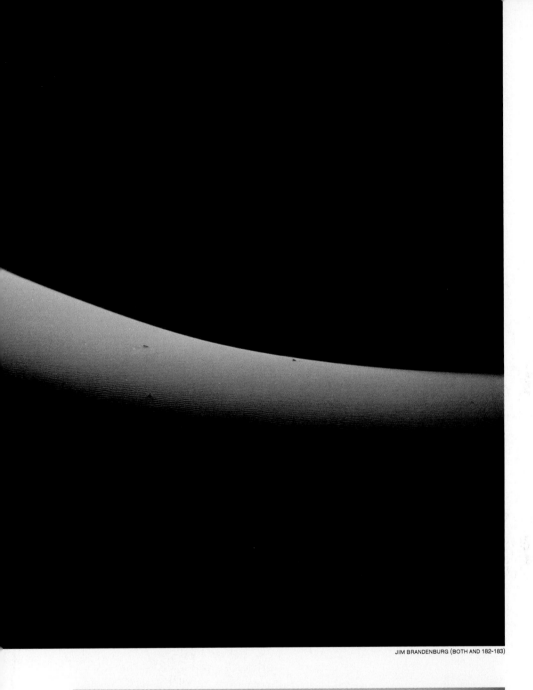

JIM BRANDENBURG (BOTH AND 182-183)

wide-reaching root network able to absorb tremendous amounts of moisture during a rainstorm, a single saguaro may store as much as six tons of water. Its accordion-like skin expands to hold the liquid load. Because of its water volume and its thick-skin insulation, this cactus has an internal temperature 20 degrees cooler than its surface temperature. Gila woodpeckers and flickers often turn the saguaro into a high-rise birdhouse by drilling holes and nesting in the cooler interior. Occasionally an elf owl—the world's smallest owl—will rout the occupants and take over the nest, which offers not only "air-conditioning" but safety high out of the reach of most predators.

Finding a cool hiding place from the sun is an excellent survival tactic in the desert. Rodents such as the kangaroo rat and the jerboa spend daylight hours in burrows where the temperature 18 inches down may be as much as 70 degrees cooler than out on the bare rock or sand. The poorwill, a nocturnal North American bird that inhabits deserts as well as forests, confounds accepted ideas about bird behavior with its adaptation. Come winter, instead of migrating to a warmer climate, the poorwill perches in a rock crevice and hibernates.

Though locating shelter is crucial for survival, the true desert imperative is conserving water. Few animals do it better than the camel. People have long believed that a camel stores water in its stomach or its hump. In truth, the camel's secret, according to Duke University physiologist Knut Schmidt-Nielsen, is that "it not only tolerates dehydration much better than a man, but also loses water much more slowly." For a year Dr. Schmidt-Nielsen studied the performance of camels in the Sahara. He found that an adult dromedary (a one-hump camel) that is deprived of water in the heat of summer may lose more than 25 percent of its body weight—or more than 200 pounds—without serious effect. By contrast, a man who loses 12 percent of his body weight in the desert will suffer from severe dehydration. He would also probably succumb to "explosive heat death," his thickened blood no longer able to expel metabolic heat from the body. The camel avoids this nasty fate by taking water from its tissues rather than from its blood.

A low rate of perspiration enables the camel to conserve water. The animal, Dr. Schmidt-Nielsen discovered, can tolerate high levels of body temperature. Its body heat can rise from 92°F all the way to 105°F before the animal begins to sweat freely. "As a result of its flexible temperature," he says, "the camel sweats little except during the hottest hours of the day. A man in the same environment, on the other hand, perspires almost from sunrise to sunset."

Joining a small camel caravan in the desert outside Tamanrasset, I learned to appreciate this ungainly beast. First of all, the view is great sitting eight feet off the ground, swaying in the slow rhythm of a boat at anchor. This was my first time aboard a camel, but except for the anxiety associated with getting on—positioning myself in the small, hard saddle at the front of the hump, having the camel lurch to its feet, and being pitched forward and back in the process—I felt comfortable. The camel seemed complacent too, as long as at the end of the day it found a thorn tree to nibble on for nourishment.

With sure footwork my camel methodically climbed up and down rough, rock-strewn ridges and plodded steadily down wadis of loose sand where Tuaregs living in thatched huts silently came out to observe our tiny caravan. The timelessness of the desert never seemed more pervasive, though, than when we rode our camels single file across a scorched, stony plain, hypnotized by the heat, the swaying, the emptiness, the only sound the murmur of our Tuareg guide as he sang gently to his camel.

Without the "ship of the desert" and its incredible endurance, the Sahara would probably have remained a blank spot on maps until the invention of the automobile. As it was, camels were introduced into North Africa in the second or first century B.C. via Egypt or Sudan. Camel caravans began penetrating the formerly inaccessible interior, establishing major trade routes between Mediterranean lands and the gold-rich kingdoms of black Africa.

Humans had taken to arid lands, however, long before they adopted camels. Irrigated desert settlements near Mount Carmel in Israel and Mount Hermon on the Lebanon-Syria border were active some 8,000 years ago. The Aboriginals of Australia's central deserts and the Kalahari's Bushmen have existed as huntergatherers for more than 20,000 years. Unlike desert Arabs who wear layers of loose robes that allow a cushion of air to circulate over their bodies, the darkskinned Aboriginals and Bushmen in some cases wear no more than loincloths. Traditionally, they survived by tracking rainstorms, walking for days to drink from a recently replenished water hole. The women and children collected edible roots and seeds, and trapped lizards. The men hunted wild game, the Bushmen with bows and arrows, the Aboriginals with spears and boomerangs.

Past desert cultures have left scant remains: a campsite strewn with bones and pottery shards, a faint ocher painting on the wall of a cave, piles of rocks for a graveyard. On a series of pebbly mesa tops in the Peruvian Desert, the most mysterious relics exist: a collection of enormous figures and geometric shapes scratched like graffiti on the desert floor. Recognized in the late 1920s, the so-called Nazca Lines are believed to have been inscribed by a pre-Inca civilization about 2,000 years ago. Markings include skillfully drawn animals such as a monkey, a hummingbird, and a spider that are hundreds of feet long.

What do the Nazca etchings signify? One unorthodox theory holds that the lines and figures were made to communicate with extraterrestrial beings. Another view describes the area as a sacred arena, the figures representing animal spirits and the lines serving as ritual pathways to various shrines. The creators left almost no clues to the meaning of their grand design.

Signs of life, past or present, soften the edges of the relentless desert. One night while camped on a plain of creosote bushes in Arizona, I stared heavenward at the distant red twinkle of Mars and shivered at the thought of a totally lifeless desert, a wilderness where dust storms may shroud the entire planet. The deserts of the earth, I began to realize then, affirm life with an intensity no other environment can match, their plants and animals (including man) defying tall odds to survive. The signs of triumph are subtle—a cactus flowering, a pair of gazelles sprinting across a sand sea, a nomad drinking from a rain puddle—but always are they inspiring. It is this poetic economy of life that, as an Arab saying bears witness, can make even the desert seem a paradise:

"The desert is the Garden of Allah, from which the Lord of the Faithful removed all superfluous human and animal life, so that there might be one place where he can walk in peace."

Amen.

*D*esire for warmth as well as modesty prompts a Bedouin girl to draw her veil about her on a cold morning near Mount Sinai, Egypt. Bedouin consider themselves citizens of the Middle Eastern deserts. Those who still follow nomadic ways pay little notice to national borders as they roam wide areas in search of forage for their sheep, goats, and camels.

PAGES 188-189: Bound for market, Daza nomads drive camels across North Africa's Sahara. World's largest and most forbidding desert, the Sahara roughly equals the United States in size.

*E*roded spires of sandstone loom
behind Tuaregs (opposite) camping on
Tassili-n-Ajjer, a mountainous plateau
in the heart of the Sahara. Now too dry
for settlement, the region once
nurtured lush grasslands that teemed
with game. A succession of cultures
called this land home. More than 9,000
years ago, inhabitants began painting
ritualistic scenes on the walls
of rock shelters that housed them.
Herdsmen who lived here between 5000
and 2000 B.C. were the most prolific
and talented artists. Two hunters (left)
stalk game. Archers (above) advance in
a detail from a rock painting depicting
a cattle raid. By the beginning
of the Christian era a climate grown
arid had transformed the hospitable
plateau into grim desert.

*W*eathered battlements of
Jaisalmer guard the Great Indian
Desert, or Thar. Ancient gazebo-
shaped monuments, foreground,
honor warriors. Above: A turban
protects a camel trader from the
blistering desert sun. The
Thar supports more people per
square mile than almost any
other parched region.
Overpopulation scalps the
scrubland and helps expand the
Thar, which now covers almost
half the northwest corner of India.

FOLLOWING PAGES: Symbol of
earth's spellbinding majesty,
Ayers Rock commands the
Australian outback. The monolith
dominates the religious
myths of Aboriginals pondering
the mysteries of creation.

ETHAN HOFFMAN/ARCHIVE (194-195)

Notes on Contributors

CHRISTINE ECKSTROM LEE majored in English at Mount Holyoke College. Since joining the Society's staff in 1974, she has coauthored *America's Atlantic Isles* and has written chapters for many books, including *Mysteries of the Ancient World, Isles of the Caribbean, Exploring America's Valleys,* and *Blue Horizons.* She also contributed to *Peoples and Places of the Past.*

During his 20 years in journalism, PAUL MARTIN has served as a military correspondent in Vietnam, edited a national boating magazine, and worked as the managing editor of a monthly medical journal. In 1979 he joined the National Geographic staff as a writer-editor for WORLD magazine. For the past five years he has been a Special Publications managing editor. He is the author of *Messengers to the Brain: Our Fantastic Five Senses,* a book for readers age 8 and older.

JANE R. McCAULEY lived three years in Geneva, Switzerland, before she became a member of the Society's staff in 1970. She is the author of two books for readers age 8 and younger—*Ways Animals Sleep* and *Baby Birds and How They Grow.* She has also contributed chapters for *Secret Corners of the World, Exploring America's Valleys,* and *America's Wild Woodlands.*

On the Society's staff since 1976, senior writer THOMAS O'NEILL has retraced the routes of pathfinder John Frémont for *Into the Wilderness,* journeyed to Pompeii for *Splendors of the Past,* and ventured into Australia's outback for *The Desert Realm.* He is the author of *Back Roads America* and *Lakes, Peaks, and Prairies: Discovering the United States-Canadian Border.*

CYNTHIA RUSS RAMSAY, a native New Yorker and a senior writer on the staff of National Geographic since 1966, has traveled widely to cover subjects related to geology, mountaineering, history, and archaeology. Her byline appears in many of the Society's books, including *Alaska's Magnificent Parklands, Splendors of the Past,* and *Nature's World of Wonders.*

Acknowledgments

The Special Publications Division is grateful to the individuals, groups, and organizations named and quoted in the text and to those cited here for their assistance during the preparation of this book:

Pierre Aases; Alice Alldredge; Barbie Allen; Jim Brandenburg; Thomas J. Casadevall; Stephen Cofer-Shabica; Mark Dimmitt; Luis Diego Gómez; Gary S. Hartshorn; Lynne F. Hartshorn; Lloyd Hulbert; David B. Lellinger; Virgil Olsen; Klaus Ruetzler; Michael P. Ryan; Aaron J. Sharp; Jeheskel Shoshani; Smithsonian Institution; U. S. Geological Survey; Masatoshi Watanabe; George E. Watson; Steven B. Young; George R. Zug.

Additional Reading

The reader may wish to consult the *National Geographic Index* for pertinent articles, and to refer to the following:

David Attenborough, *The Living Planet;* Fred M. Bullard, *Volcanoes of the Earth;* Catherine Caufield, *In the Rainforest;* David F. Costello, *The Prairie World;* Robert and Barbara Decker, *Volcano Watching;* Iain and Oria Douglas-Hamilton, *Among the Elephants;* Uwe George, *In the Deserts of the Earth;* Georg Gerster, *Sahara;* Bernhard and Michel Grzimek, *Serengeti Shall Not Die;* Jeremy Keenan, *The Tuareg;* Peggy Larson, *The Deserts of the Southwest;* J. D. Love and John C. Reed, Jr., *Creation of the Teton Landscape;* F. A. McClure, *The Bamboos;* Tony Morrison, *Land Above the Clouds;* Cynthia Moss, *Portraits in the Wild;* John Muir, *Travels in Alaska;* Jack Page, *Arid Lands;* Paul W. Richards, *The Life of the Jungle;* Robert Silverberg, *The World of the Rain Forest;* Simon and Schuster, *The International Book of the Forest;* Ann Sutton and Myron Sutton, *Eastern Forests;* Jeremy Swift, *The Sahara.*

Tailored for a sandstone giant, the Mittens reach a thousand feet above Monument Valley on the Arizona-Utah border. Early day prospectors who gave the buttes their name thought they resembled mittened hands.

Index

Library of Congress CIP Data
Main entry under title:

Our awesome earth.

  Bibliography: p.
  Includes index.
  1. Natural history. 2. Earth sciences. I.
National Geographic Society (U.S.).
Special Publications Division.
QH45.5.O87   1986   508   85-23733
ISBN 0-87044-545-6 (regular edition)
ISBN 0-87044-550-2 (library edition)

Composition for *Our Awesome Earth: Its Mysteries and Its Splendors* by National Geographic's Photographic Services, Carl M. Shrader, Director, Lawrence F. Ludwig, Assistant Director. Printed and bound by Holladay-Tyler Printing Corp., Rockville, Md. Film preparation by Catharine Cooke Studio, Inc., New York, N.Y. Color separations by the Lanman Progressive Company, Washington, D.C.; Lincoln Graphics, Inc., Cherry Hill, N.J.; and NEC, Inc., Nashville, Tenn.